HANDBOOK of LABORATORY ANIMAL BACTERIOLOGY

HANDBOOK of LABORATORY ANIMAL BACTERIOLOGY

Axel Kornerup Hansen

CRC Press
Boca Raton   London   New York   Washington, D.C.

Contact Editor: John Sulzycki
Project Editor: Christine Andreasen
Cover Design: Dawn Boyd

Library of Congress Cataloging-in-Publication Data

Hansen, Axel Kornerup.
    Handbook of laboratory animal bacteriology / Axel Kornerup Hansen.
        p. cm.
    Includes bibliographical references and index.
    ISBN 0-8493-2913-2
    1. Laboratory animals--Diseases--Diagnosis. 2. Laboratory animals--Health. 3. Veterinary bacteriology--Technique. I. Title.
    SF996.5.H35 1999
    636.088'5—dc21

                                        99-23333
                                              CIP

No claim to original U.S. Government works
International Standard Book Number 0-8493-2913-2
Library of Congress Card Number 99-23333
Printed in the United States of America   2 3 4 5 6 7 8 9 0
Printed on acid-free paper

Preface

Bacteriological examination of laboratory animals has become an important part of ensuring the quality of these animals for research purposes. During the last decade, sets of international guidelines have been published to force vendors and institutions to perform health monitoring, including bacteriological screenings, on their animals. Such guidelines, however, only rarely provide any detailed advice on how to perform bacteriological procedures on laboratory animals.

Bacteriological procedures for examining healthy laboratory animals often differ from diagnostic procedures for diseased animals, and therefore laboratory animal bacteriologists often perform procedures that may not be found in ordinary textbooks on bacteriology. The bacterial species dealt with also often differ greatly from species covered by textbooks on veterinary or human bacteriology. It is, therefore, the scope of this book to provide the reader with such information. It is my hope that this book will be on the shelves of laboratory animal health monitoring laboratories and that information I myself have often had to search for in a range of sources during my career will now be readily at hand. I also hope that the book will be useful to those who, within a short time, need to acquire overall information on how to perform laboratory animal bacteriological procedures. Finally, I hope the book will inspire more laboratories to offer bacteriological examinations of laboratory animals and thereby take part in the global efforts to raise the quality of laboratory animals, for the sake of humans and for the sake of the animals.

The first part of the book provides information on how to sample and cultivate from animals. General descriptions of various identification procedures follow, and finally important laboratory animal bacteria are described. Although I recognize that a variety of animal species are used as laboratory animals, the latter sections are limited to providing information on bacteria isolated from rodents and rabbits, since other animal species are covered in textbooks on veterinary bacteriology.

I wish to express my sincere thanks to all those laboratory technicians who, over the years, have helped me to perform all these procedures. I especially want to thank my wife, Inge, for encouraging and helping me and for taking care of a lot of everyday practical matters while I was writing this book, and my children, Rikke and Emil, who kept themselves quiet long enough to allow me to complete this work.

Axel Kornerup Hansen
Professor, dr.vet.sci., DVM
Copenhagen

About the Author

Axel Kornerup Hansen, dr.vet. sci., DVM, is Professor of Laboratory Animal Science and Welfare at the Royal Veterinary and Agricultural University in Copenhagen, Denmark, from which he also graduated in veterinary medicine in 1985. In 1987 after a period in small animal practices he became head of the health monitoring laboratory at Møllegaard Breeding Centre (now M&B Ltd.), where his activities included the study of the epidemiology and research interference of laboratory animal bacteria. In 1993 he became Assistant Director of the Department of Experimental Medicine at the University of Copenhagen, where he was in charge of a major health monitoring laboratory performing a range of examinations in bacteriology, virology, and parasitology. In 1996 he earned his doctorate with a thesis on health monitoring and microbiological quality of laboratory rats. Since 1997 he has been Professor, responsible for teaching undergraduate and postgraduate students of laboratory animal science.

Dr. Hansen is a board member of the Scandinavian Society for Laboratory Animal Science and the Federation of European Laboratory Animal Science Associations; national editor of *The Scandinavian Journal of Laboratory Animal Science;* steering group member of the European College of Laboratory Animal Medicine; and a member of AALAS. He has organized several international meetings and lectures on laboratory animal science in Denmark, Estonia, Germany, Korea, Latvia, Lithuania, Norway, Russia, Sweden, the United Kingdom, and the United States.

Dr. Hansen has published more than 100 scientific papers and chapters in textbooks. He has been involved in setting up European guidelines for health monitoring of a range of animal species, as well as standards for accreditation of health monitoring laboratories.

Dr. Hansen's current research interests include laboratory animal infections and methods for improving the welfare of animals used for research.

Contents

1. **Strategies for sampling animals
   for bacteriological examination** 1
   1.1 Bacterial interference with research 1
   1.2 Bacteriological examination of healthy animals 5
       1.2.1 The aim of examining healthy animals..............................5
       1.2.2 Sampling strategies 6
           1.2.2.1 Random sampling 6
           1.2.2.2 Calculation of the sample size 7
           1.2.2.3 Sampling frequency 9
           1.2.2.4 Defining the microbiological entity 10
       1.2.3 Immunosuppression of animals prior to sampling
               (stress testing) 10
       1.2.4 Screening vs. profiling..11
   1.3 Bacteriological examination of diseased animals 11
   References...12

2. **Sampling procedures**...15
   2.1 Planning the work 15
   2.2 Euthanasia 15
   2.3 Blood sampling...18
   2.4 Instruments and sterilizing procedures during sampling 20
   2.5 Opening and inspecting the carcass 21
   2.6 Sampling from various organs 21
   References...46

3. **Cultivation methods**...47
   3.1 The choice of media..47
   3.2 Incubation of media..49
   3.3 Isolation of bacteria 51
   References...54

4. **Identification of bacteria** 55
   4.1 Initial characterization of the isolates..55
   4.2 Conclusive identification. 56

4.3   Specific techniques used for identification of bacteria 60
      4.3.1   Gram-stainability tests 60
      4.3.2   Other methods used for describing
              the shape of bacteria 60
      4.3.3   Motility tests 62
      4.3.4   Test for aerobic and anaerobic growth 62
      4.3.5   Catalase test 63
      4.3.6   Cytochrome oxidase test 64
      4.3.7   Acid-fast or spore staining 64
      4.3.8   Carbohydrate fermentation and utilization assays 64
      4.3.9   Disc methods 65
      4.3.10 Other assays 68
      4.3.11 Commercial test kits 68
References 71

5. **Immunological methods** **75**
    5.1   Agglutination 76
    5.2   Immunofluorescence techniques 77
          5.2.1   Diagnosing the presence of bacteria in a sample 77
          5.2.2   The immunofluorescence assay 81
    5.3   Immunoenzymatic staining 82
    5.4   Enzyme-linked immunosorbent assay 86
          5.4.1   Principles 86
          5.4.2   The microtiter plates 87
          5.4.3   The antigen 93
          5.4.4   Antibodies, enzymes, and substrates 93
          5.4.5   Coating the wells 94
          5.4.6   Blocking the wells 94
          5.4.7   Performing the assay 95
          5.4.8   Control sera 95
          5.4.9   Interpretation of the OD-value 97
    5.5   Immunoblotting 98
    References 99

6. **Molecular biological methods** **101**
    6.1   Detection by molecular probes 105
          6.1.1   Solid-phase hybridization 105
          6.1.2   Solution-phase hybridization 105
    6.2   Amplification of nucleic acids 105
          6.2.1   Principle 105
          6.2.2   The polymerase enzyme 108
          6.2.3   Reverse transcriptase RNA PCR 108
          6.2.4   The template sequence and the primers 109
          6.2.5   Preparation of the sample 109
          6.2.6   Detection of the PCR product 109

     6.2.7   Contamination of the amplification process 110

  6.3   Restriction analysis of chromosomal DNA 110

References.. 111

7. Gram-positive cocci **113**

  7.1   Micrococcaceae 114

     7.1.1   *Micrococcus* 114

     7.1.2   *Staphylococcus* 114

         7.1.2.1   Characteristics of infection 114

         7.1.2.2   Characteristics of the agent 116

             7.1.2.2.1   Morphology 116

             7.1.2.2.2   Cultivation 117

             7.1.2.2.3   Isolation sites 118

             7.1.2.2.4   Differentiation and identification 118

  7.2   Streptococcaceae 118

     7.2.1   *Streptococcus* 118

         7.2.1.1   Characteristics of the infection 118

             7.2.1.1.1   Epidemiology... 118

             7.2.1.1.2   Clinical disease and pathology........... 120

             7.2.1.1.3   Interference with research 121

         7.2.1.2   Characteristics of the agent 121

             7.2.1.2.1   Morphology 121

             7.2.1.2.2   Media 121

             7.2.1.2.3   Sampling sites.. 121

             7.2.1.2.4   Immunological identification

                      of isolates.. 121

             7.2.1.2.5   Biochemical identification

                      of isolates.. 122

     7.2.2   *Enterococcus* 123

     7.2.3   *Aerococcus*... 123

     7.2.4   *Gemella*.. 123

  7.3   Peptococcaceae 124

References.. 124

8. Gram-positive rods isolated by simple cultivation............................ **127**

  8.1   Regular nonacid-fast, nonsporogenic rods 128

     8.1.1   Obligate aerobic regular nonsporogenic rods 128

     8.1.2   Facultatively anaerobic regular nonsporogenic rods 130

         8.1.2.1   *Lactobacillus*.. 130

         8.1.2.2   *Erysipelothrix*.. 131

             8.1.2.2.1   Characteristics of infection................... 131

             8.1.2.2.2   Characteristics of the agent................. 131

         8.1.2.3   *Listeria*... 132

             8.1.2.3.1   Characteristics of infection................... 132

             8.1.2.3.2   Characteristics of the agent................. 133

8.2   Coryneform bacteria..134
    8.2.1   *Corynebacterium* spp..134
        8.2.1.1 Characteristics of infection 134
        8.2.1.2 Characteristics of the agent 137
            8.2.1.2.1 Morphology 137
            8.2.1.2.2 Cultivation 138
            8.2.1.2.3 Isolation sites 138
            8.2.1.2.4 Differentiation and identification.......138
    8.2.2   *Arcanobacterium*...138
    8.2.3   *Rhodococcus* 139
    8.2.4   *Oerskovia* 139
    8.2.5   *Eubacterium* 139
8.3   Gram-positive rods forming mycelium or hyphae............140
    8.3.1   *Actinomyces* spp. 140
        8.3.1.1 Characteristics of infection 140
        8.3.1.2 Characteristics of the agent 140
8.4   Gram-positive sporogenic rods 141
    8.4.1   *Bacillus*...141
        8.4.1.1 Characteristics of infection 141
        8.4.1.2 Characteristics of the agent 142
    8.4.2   *Clostridium* 143
        8.4.2.1 Characteristics of infection 143
        8.4.2.2 Characteristics of the agent 145
            8.4.2.2.1 Morphology 145
            8.4.2.2.2 Cultivation 146
            8.4.2.2.3 Isolation sites 146
            8.4.2.2.4 Differentiation and identification.......146
            8.4.2.2.5 Toxin detection 147
References...150

9.   **Gram-positive rods not cultivated by simple techniques**..............153
9.1   *Mycobacterium* spp...154
    9.1.1   Characteristics of infection 154
    9.1.2   Characteristics of the agent 154
        9.1.2.1 Morphology 154
        9.1.2.2 Cultivation..154
        9.1.2.3 Isolation sites 156
        9.1.2.4 Differentiation and identification 156
    9.1.3   Safety..156
9.2   *Clostridium piliforme* 158
    9.2.1   Characteristics of infection 158
    9.2.2   Characteristics of the agent 159
        9.2.2.1 Morphology..159

        9.2.2.2 Cultivation...159
        9.2.2.3 Staining..160
        9.2.2.4 Differentiation and identification........................160
        9.2.2.5 Serology...160
        9.2.2.6 Molecular biology...160
    References..161

10. **Facultatively anaerobic Gram-negative bacteria................................165**
    10.1 Enterobacteriaceae...166
        10.1.1 Characteristics of infection.......................................166
            10.1.1.1 *Escherichia*..167
            10.1.1.2 *Citrobacter*...167
            10.1.1.3 *Salmonella*..168
            10.1.1.4 *Klebsiella*..168
            10.1.1.5 *Enterobacter*..168
            10.1.1.6 *Proteus*..168
            10.1.1.7 *Morganella*..168
            10.1.1.8 *Yersinia*...169
            10.1.1.9 Other species...169
        10.1.2 Characteristics of the agent.....................................169
            10.1.2.1. Sampling and cultivation...........................169
            10.1.2.2 Identification...171
        10.1.3 Safety...176
    10.2 Pasteurellaceae...177
        10.2.1 *Pasteurella*..177
            10.2.1.1 Characteristics of infection.......................177
            10.2.1.2 Characteristics of the agent......................177
                10.2.1.2.1 Morphology.....................................177
                10.2.1.2.2 Cultivation.......................................178
                10.2.1.2.3 Isolation sites...................................178
                10.2.1.2.4 Differentiation and identification.......178
                10.2.1.2.5 Serology...178
        10.2.2 *Actinobacillus*...179
            10.2.2.1 Characteristics of infection.......................179
            10.2.2.2 Characteristics of the agent......................180
                10.2.2.2.1 Morphology.....................................180
                10.2.2.2.2 Cultivation.......................................180
                10.2.2.2.3 Isolation sites...................................180
                10.2.2.2.4 Differentiation and identification.......180
                10.2.2.2.5 Serology...181
                10.2.2.2.6 Molecular biology............................181
        10.2.3 *Haemophilus*...181
            10.2.3.1 Characteristics of infection.......................181
            10.2.3.2 Characteristics of the agent......................181

10.3 *Streptobacillus moniliformis* 182
    10.3.1 Characteristics of infection 182
    10.3.2 Characteristics of the agent 184
        10.3.2.1 Morphology 184
        10.3.2.2 Cultivation...184
        10.3.2.3 Isolation sites 184
        10.3.2.4 Differentiation and identification 184
        10.3.2.5 Serology 184
10.4 Vibrionaceae...185
References..186

11. **Obligate aerobic, Gram-negative rods** **191**
  11.1 Motile, aerobic, Gram-negative rods 192
    11.1.1 *Pseudomonas*...192
        11.1.1.1 Characteristics of infection 192
        11.1.1.2 Characteristics of the agent 194
            11.1.1.2.1 Morphology 194
            11.1.1.2.2 Cultivation 195
            11.1.1.2.3 Isolation sites 195
            11.1.1.2.4 Differentiation and identification.......195
            11.1.1.2.5 Serology.................................195
    11.1.2 *Xanthomonas, Sphingomonas, Chryseomonas,*
        Agrobacterium, and *Burkholderia* 195
    11.1.3 *Bordetella* 197
        11.1.3.1 Characteristics of infection 197
        11.1.3.2 Characteristics of the agent 197
            11.1.3.2.1 Morphology 197
            11.1.3.2.2 Cultivation 197
            11.1.3.2.3 Isolation sites 197
            11.1.3.2.4 Differentiation and identification.......197
            11.1.3.2.5 Serology.................................197
  11.2 Non-motile, aerobic, Gram-negative rods that can
    easily be cultivated 198
    11.2.1 *Acinetobacter* 198
    11.2.2 *Flavobacterium*...198
    11.2.3 *Weeksella* 198
  11.3 Non-motile, aerobic, Gram-negative rods that
    are difficult to cultivate..198
    11.3.1 *Francisella* 200
        11.3.1.1 Characteristics of infection 200
        11.3.1.2 Characteristics of the agent 200
        11.3.1.3 Safety...200
  References..201

12. **Obligate anaerobic and microaerophilic Gram-negative bacteria** 203
   12.1 Bacteroidaceae 204
      12.1.1 *Fusobacterium* 204
         12.1.1.1 Characteristics of infection 204
         12.1.1.2 Characteristics of the agent 204
            12.1.1.2.1 Morphology 204
            12.1.1.2.2 Cultivation 206
            12.1.1.2.3 Differentiation and identification 206
   12.2 Microaerophilic curved bacteria 206
      12.2.1 *Campylobacter* 206
         12.2.1.1 Characteristics of infection 206
         12.2.1.2 Characteristics of the agent 208
            12.2.1.2.1 Morphology 208
            12.2.1.2.2 Cultivation 208
            12.2.1.2.3 Differentiation and identification 208
         12.2.1.3 Safety 209
      12.2.2 *Helicobacter* 209
         12.2.2.1 Characteristics of infection 209
         12.2.2.2 Characteristics of the agent 210
            12.2.2.2.1 Morphology 210
            12.2.2.2.2 Isolation sites and cultivation 210
            12.2.2.2.3 Differentiation and identification 210
            12.2.2.2.4 Serology 211
            12.2.2.2.5 Molecular biology 211
   12.3 Cilia-associated respiratory bacillus 211
      12.3.1 Characteristics of infection 211
      12.3.2 Characteristics of the agent 212
         12.3.2.1 Isolation 212
         12.3.2.2 Identification 212
   References 212

13. **Spirochetes** 215
   13.1 *Treponema* 216
      13.1.1 Characteristics of infection 216
      13.1.2 Characteristics of the agent 216
         13.1.2.1 Morphology 216
         13.1.2.2 Microscopy 216
         13.1.2.3 Cultivation 217
         13.1.2.4 Serology 217
         13.1.2.5 Molecular biology 217
   13.2 *Leptospira* 218
      13.2.1 Characteristics of infection 218

13.2.2 Characteristics of the agent 218
        13.2.2.1 Microscopy 218
        13.2.2.2 Cultivation.. 219
        13.2.2.3 Differentiation and identification 219
        13.2.2.4 Serology 220
13.2.3 Safety 220
13.3 *Spirillum minus* 220
References 220

14. Mollicutes 223
14.1 *Mycoplasma* 223
14.1.1 Characteristics of infection 223
14.1.2 Characteristics of the agent 225
        14.1.2.1 Morphology 225
        14.1.2.2 Cultivation.. 225
        14.1.2.3 Identification 228
        14.1.2.4 Serology 228
14.2 *Acholeplasma* 229
References 229

Appendix I Producers of Reagents for Laboratory
Animal Bacteriology 233

Appendix II Biosafety Levels for Microbiological Laboratories 239

Index 245

chapter one

Strategies for sampling animals for bacteriological examination

Contents

1.1   Bacterial interference with research..1
1.2   Bacteriological examination of healthy animals.........................5
    1.2.1   The aim of examining healthy animals 5
    1.2.2   Sampling strategies 6
        1.2.2.1   Random sampling 6
        1.2.2.2   Calculation of sample size 7
        1.2.2.3   Sampling frequency 9
        1.2.2.4   Defining the microbiological entity.............10
    1.2.3   Immunosuppression of animals prior to sampling
        (stress testing)..10
    1.2.4   Screening vs. profiling 11
1.3   Bacteriological examination of diseased animals....................11
References..12

1.1 Bacterial interference with research

In veterinary medicine, the traditional reason for showing interest in infectious agents has been their ability to cause disease, i.e., their *patho-genicity*. This is, of course, also vital in laboratory animal medicine, as only healthy animals should be used for experiments. However, the impact of infections goes far beyond this, as infections may interfere with research, even in the absence of clinical symptoms or pathological changes. Infections may have a direct impact on the immunology, the physiology, the reproduction, other parts of the microbiology, and the oncology.[1] Even if the study includes a test group and a control group, infections are still important because microorganisms:

- May inhibit the induction of a certain animal model
- May make it difficult to interpret the final results
- May be dose-related in their effect, if the test factor acts as an inducer
- May increase the variation inside the group, thereby leading to the use of a larger number of animals

The interference may also be more indirect, e.g., by the contamination of biological products used for experiments.

In a very broad sense, the word *infection* is used within laboratory animal medicine to cover the presence of a microorganism — *the agent* — in both an animal host and in a colony including several hosts. As the aim of performing bacteriological examinations on laboratory animals is more in the direction of preventing infectious impact rather than diagnosing it after it has already occurred, the status of the colony in most cases is more important than the status of the individual.

Terms used to describe infections on different levels are given in Table 1.1. Bacteria differ from viruses and parasites in the sense that all conventional and barrier-bred animals do harbor a range of different bacteria, while animals can be kept free of both viruses and parasites. This is, however, impossible for bacteria unless animals are reared in isolators. Bacterial impact on experiments is far less controllable and definable than the impact caused by other types of microorganisms.

Some microorganisms only influence the animal temporarily, while others — and that certainly goes for most bacteria — act over a long period, perhaps even for the entire life of the animal. This division is mainly connected with the ability of the agent to persist in the organism. This is, however, not always so, as can be illustrated by the very simple example given by those infections that induce resistance against reinfection. The agent leaves the organism, having caused a permanent change in the immune system. Such examples also illustrate that a microorganism interferes with a certain part of the animal permanently, in this case the immune system, while other parts of the animal, e.g., its anatomy, are only affected temporarily. If an animal colony is infected with microorganisms that only temporarily influence the individuals, this influence may only be observed in animals of a certain age, and many studies will not be influenced. Other studies may incidentally be designed in such a way that they use exactly those animals interfered with.

The impact on research caused by a bacterial infection is not solely due to factors of the microorganism. Also of importance are host genetics, the environment, and the experiment itself, as well as the balance between these determinants.

Apart from having many nonmicrobial side effects on the animals, environmental stress may induce infectious disease, e.g., raised air concentrations of NH_3 or deficiency of vitamins A and E may induce respiratory disease in rats latently infected with *Mycoplasma pulmonis*.[2-5] Acidification of the drinking water is known to reduce the prevalence of *Pseudomonas aeruginosa*

Table 1.1 Terms Used in Laboratory Animal Epizootiology

Agent	The causative organism in a specific infection
Association	The presence of a symbiont or commensal in an animal host
Commensal	Microorganism that has no impact on the host organism
Copathogen	A microorganism-enhancing disease primarily produced by another agent
Determinants	All factors, the agent included, involved in the development of disease
Disease	Clinical manifestation of infection
Monofactorial disease	Disease in which the agent is the only factor influencing whether disease is observed or not
Multifactorial disease	Disease which is the result of several factors, including the agent
Health status	The actual status of an individual animal concerning its clinical, pathological, and physiological appearance
Inapparent carrier	The animal harboring a subclinical infection
Incidence	The fraction of new cases of infection over a defined period
Infection	The presence of a microorganism in an animal host or in an animal colony
Subclinical Infection	Infection without clinical manifestations
Dormant subclinical infection	Subclinical infection in which the agent can be recovered
Latent subclinical infection	Subclinical infection in which the presence of the agent can only be proved by indirect methods
Inducer	Determinant enhancing disease in the animal
Infectivity	The ability of the microorganism to infect individuals
Opportunist	Facultative pathogen
Pathogen	A microorganism capable of inducing disease
Obligate pathogen	Microorganism always causing disease
Facultative pathogen	Microorganism only causing disease in connection with inducers
Pathogenicity	The ability of a certain microorganism to cause disease
Pathogenic	Capable of causing disease
Prevalence	The fraction of animals infected with a certain microorganism at a certain point of time
Protector	Determinant protecting against disease in the animal
Saprophytic	Incapable of causing disease
Symbiont	Microorganism which has a positive value for the host organism
Virulence	A quantitative measure of the pathogenicity of the agent

significantly.[6] Deficiency of vitamin C in guinea pigs may be responsible for a high incidence of bacterial diseases caused by *Streptococcus* or *Klebsiella* spp.,[7] for example, while iron deficiency has been described as an inducer of respiratory disease caused by *S. pneumoniae* in rats.[8] Several nutritional components work as determinants for the development of Tyzzer's disease in mice infected with *Clostridium piliforme*, as high contents of glucose and peanut oil seem to protect against disease, while a high content of casein induces disease.[9] The proper ventilation of the animal unit is essential. Lowered air exchange may lead to respiratory disease caused by bacteria with low pathogenicity such as *Staphylococcus xylosus* and *P. aeruginosa*, especially in immune-deficient animals.[10]

Animal susceptibility to the development of disease is controlled by genetics, sex, sexual cycle, age, and other characteristics of the host. For example, immune-deficient animals are obviously more susceptible to the development of infectious disease, such as that frequently experienced in the nude mouse, which may suffer from abscesses caused by bacterial species such as *Staphylococcus aureus*, *Pasteurella pneumotropica*, *Morganella morganii*, *Citrobacter* spp., and *Streptococcus* spp.[11–13] Inbred strains of rats are known to show variation in susceptibility to the development of mycoplasmosis[14] and Tyzzer's disease.[15] Bacterial disease, e.g., Tyzzer's disease,[16–17] is most common shortly after weaning, as these animals are stressed by their new environment and are no longer protected by maternal antibodies. However, if the animal is infected with the agent for the first time late in life, or a persistent infection is reactivated by stress or immunosuppression, disease is often worse than disease observed in younger animals, e.g., in the case of mycoplasmosis.[18] Infection with *group C Streptococcus* in guinea pigs is more common in females,[19] while colitis and rectal prolapse caused by *Citrobacter rodentium* in mice are more common in males and might be prevented by castration.[20]

Virulent genes of different bacteria code for differences in virulence. The application of good hygienic principles will reduce the number of animals exposed to a lethal dose of bacteria, as there often is a gap between the infective dose and the lethal dose. For example, in guinea pigs the infective dose of *Bordetella bronchiseptica* has been found to be 4 colony-forming units, while the lethal dose has been found to be 1314 colony-forming units.[21] The varying degrees of susceptibility to the development of Tyzzer's disease observed in different animal species may be connected with differences in the pathogenicity of the strains of *C. piliforme* infecting these species.[22–24] Some bacteria enhance the impact of other bacteria, e.g., respiratory disease as a cause of *M. pulmonis* will be enhanced by the presence of CAR bacillus or *P. pneumotropica*.[26–27] Some bacteria, e.g., *Streptococcus* spp. and *Lactobacillus* spp., have a probiotic effect, i.e., they reduce the incidence of enteric bacterial disease caused by *Escherichia coli*, for example.

The experiment itself may be a stress factor. Postsurgical infections with bacteria such as *P. aeruginosa, E. coli,* and *S. aureus* are well known.[28,29] A special problem in laboratory animal medicine is imposed by experimental

Table 1.2   Examples of Latent Bacterial Infections Which May Be Activated
by Experimental Immunosuppression

Organism	Effect	Species
Citrobacter rodentium	Ulcers in duodenum and ileum, enteritis	Mouse
Clostridium piliforme	Tyzzer's disease	Mouse, rat, hamster
Corynebacterium kutscheri	Pseudotuberculosis	Mouse
Enterobacter cloacae	Fatal neonatal diarrhea, death	Mouse
Klebsiella pneumoniae	Death	Mouse
Pseudomonas aeruginosa	Bacteremia, liver necrosis, lowered body weight	Mouse
Salmonella spp.	Enteritis	Rat
Staphylococcus spp.	Respiratory disease	Rat
Streptobacillus moniliformis	Splenomegaly	Mouse

Data from Hansen, A. K., *Handbook of Laboratory Animal Science, Vol. I, Animal Selection and Handling in Biomedical Research,* Svendsen, P. and Hau, J., Eds., CRC Press, Boca Raton, 1994, chap. 11.

immunosuppression, which is a common tool in many experiments. Various bacteria, even members of the normal flora, cause disease and death in immunosuppressed animals (Table 1.2). A potent immunosuppressor may turn almost any apathogen bacteria into a pathogen, and if such experiments are to be performed on animals that are not gnotobiotic, at least good hygiene is essential. The induction of specific disease models often stresses the animals, thereby making them more susceptible to disease. For example, the induction of pulmonary edema is known to activate infections with *S. pneumoniae* in rats, resulting in different types of lung disease.[30]

1.2   Bacteriological examination of healthy animals

1.2.1   The aim of examining healthy animals

Due to what has been described above it is essential that rodents used for biomedical research are *microbiologically defined*, that is, that the microbiological status of the animals used is known. However, as evaluation of the microbiology of the individual animal taking part in the experiment is difficult, these factors generally will be monitored in animals sampled only for this purpose, and status will be a picture of the population and not the individual. For example, a certain rat colony may be infected with *Salmonella*, which means that there is a certain risk that rats from this colony may develop salmonellosis during the experiment. However, it does not mean that the rats in the experiment necessarily carry *Salmonella* and that they will develop salmonellosis, only that there is a certain risk that it will be so. Thus, the microbiological status is generally given as a list of microorganisms monitored within the colony, with a designation as to whether or not they have been found, without any remarks as to the organisms causing disease.

Defining the microbiological status is usually done by the routine sampling of healthy animals from the colony. To reveal bacteria with a certain research-interfering potential samples are taken from these animals. The bacteria are normally not in a state of causing disease in that animal and are therefore not found as pure cultures in the animal. Therefore, bacteria of interest have to be isolated among many other bacteria (of no interest) also found in the animal.

How many animals to sample, how often to do it, what to look for, and which methods to choose should be based on scientific judgment dependent on the type of research and animals involved, but international guidelines may be consulted. In Europe a set of recommendations has been issued by international organizations within laboratory animal science, e.g., the *Federation of European Laboratory Animal Science Associations (FELASA)*[31,32] or the *Scandinavian Society for Laboratory Animal Science (Scand-LAS)*.[33] Table 1.3 shows which bacteria these guidelines recommend for routine health monitoring. The guidelines give only very sparse information on methodology for isolating bacteria. The scope of this book is to provide precise details of how this kind of examination may be performed. As an alternative to the FELASA guidelines, the "Manual of Microbiologic Monitoring of Laboratory Animals"[34] from the U.S. National Institutes of Health may be consulted.

1.2.2   Sampling strategies*

1.2.2.1   Random sampling

Health monitoring of laboratory animal colonies is based on three basic principles:

1. A few animals can be sampled for examination, but the results can be used to describe the entire colony.
2. If one animal is found to be infected with a certain organism the entire colony is considered infected with that particular organism.
3. If no animals are found to be infected with a certain organism the entire colony is considered free of that particular organism.

Principle 1 presupposes independence between the animals sampled. One can define the group to sample from, e.g., it is known that *C. piliforme* is usually only isolated from rats if these rats are examined shortly after weaning,[35] i.e., all animals sampled should be of that age. *Sensu strictu*, the

---

* Portions of this section are from Hansen, A.K., *Scand. J. Lab. Anim. Sci.*, 20(1), 11, 1993. With permission.

Table 1.3   Bacteria To Be Included in Health Monitoring
Programs for the Animals Listed[a]

	Mice	Rats	Hamsters	Guinea pigs	Gerbils	Rabbits
Bordetella bronchiseptica	+	+	+	+	+	+
CAR bacillus*	–	+	–	–	–	+
Citrobacter rodentium	+	–	–	–	–	–
Clostridium piliforme	+	+	+	+	+	+
Corynebacterium kutscheri	+	+	–	–	–	–
Helicobacter spp.*	+	–	–	–	–	–
Leptospira spp.	+	+	–	–	–	–
Mycoplasma spp.	+	+	–	–	–	–
Pasteurella spp.	+	+	+	+	+	+
Salmonella	+	+	+	+	+	+
Streptobacillus moniliformis	+	+	–	+	–	–
β-Hemolytic streptococci	+	+	–	+	+	+
Streptococcus pneumoniae	+	+	–	+	+	–
Yersinia pseudotuberculosis	–	–	–	+	–	–

Note: The agents noted with * either are not listed or are not mandatory in the guidelines,
 although it may be recommended to include them as well.

[a] According to international guidelines for health monitoring.[31–33]

results will only be valid for animals of that age, but it is assumed that if the agent is not found in animals of the specified age it is not to be found at all. When one or several of such criteria have been made out, it is of vital importance to sample among the animals fulfilling these criteria in a way that avoids the influence of other criteria, e.g., animals must not be sampled from the same cage, from the same end of the unit, etc. If such independence claims are not fulfilled, the conclusion cannot be extended to cover all animals within the unit.

1.2.2.2   Calculation of the sample size

The fraction of animals in a colony infected at a certain moment is termed the *instantaneous prevalence rate* or simply the *prevalence* (p).[36] The prevalence that a certain infection reaches depends on many factors, e.g., the contact between the animals, the resistance of the animals, etc. However, characteristics of the agent itself play a major role. It is the experience in laboratory animal epidemiology that the observed prevalences of a certain agent usually

8 Handbook of Laboratory Animal Bacteriology

fall within a certain range.[37] If sampling itself does not affect the prevalence, it can, therefore, be assumed that the prevalence is independent of the population size. If all the infected animals, and only the infected animals, in a population react positively in a given test system, then the risk of reaching a false-positive diagnosis (C) by sampling one animal is

$$C = 1 - p \tag{1}$$

while the sampling of S animals gives

$$C = (1 - p)^s \tag{2}$$

in which S is the sample size. This gives the equation for the sample sizes normally used for health monitoring in colonies of laboratory animals:[38]

$$S \geq \frac{\log C}{\log(1 - p)} \tag{3}$$

If the examination is to have more than 95% probability of being correct *the confidence limit C* is 0.05. If certain criteria, as described in Section 1.2.2.1, are outlined prior to sampling, the prevalence to be used is the prevalence in the group sampled from, and not the overall prevalence in the colony.

Equation 3 is, however, too simplified. When using the term prevalence it is assumed that this is *the real prevalence*, which means the number of animals actually infected with the organism. However, in reality it does not matter how many animals are infected, only how many react positively in the test system used. For example, if no infected animals react positively in the system, then the chance of reaching a false diagnosis would be 100%.

Table 1.4 Definition of the Terms
(a) True Positives; (b) False Negatives;
(c) False Positives; (d) True Negatives

	Test result	
	Positive	Negative
Infected	a	b
Not infected	c	d

Based on Table 1.4 the following equations can be made:[39,40]

$$\textit{Nosografic sensitivity } (N_1) = \frac{a}{a + b} \quad \text{Fraction of infected animals reacting positively in the test} \tag{4}$$

$$Nosographic\ specificity\ (N_2)\ =\ \frac{d}{c+d} \quad \begin{array}{l}\text{Fraction of non-infected animals} \\ \text{reacting negatively in the test}\end{array} \quad (5)$$

$$Diagnostic\ sensitivity\ (D_1)\ =\ \frac{d}{b+d} \quad \begin{array}{l}\text{Fraction of negative results} \\ \text{caused by non-infected animals}\end{array} \quad (6)$$

$$Diagnostic\ specificity\ (D_2)\ =\ \frac{a}{a+c} \quad \begin{array}{l}\text{Fraction of positive results caused} \\ \text{by infected animals}\end{array} \quad (7)$$

The prevalence to be used for calculation of the sample size is only that fraction out of the real prevalence p, which also will react positively in the test, i.e., $p \cdot N_1$, which leads to the following equation:[41]

$$S \geq \frac{\log C}{\log(1 - (p \cdot N_1))} \qquad (8)$$

This relationship is shown in Figure 1.1.

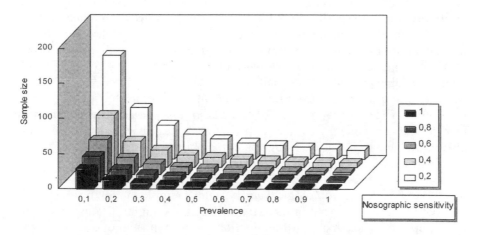

Figure 1.1 The correlation between sample size and expected prevalence of a certain infection at different nosographic sensitivities. The risk of a false-negative result is 0.05 for all bars.

1.2.2.3 Sampling frequency

A sample visualizes the status at the moment of sampling. As soon as it has been taken, it becomes historical. Only curiosity will dictate when to take the next sample. If two samplings taken at a certain interval are compared the last sampling will visualize whether changes have occurred between the two samplings. It could be argued that if 50 animals must be sampled to reach a confidence limit above 95%, and one animal is sampled every week, the confident sample size is reached after a year. The calculation may be correct, at least for persisting infections, but what has been calculated is a 95% confidence of the status a year ago.

1.2.2.4   *Defining the microbiological entity*

It is rather obvious that a random sample of animals may only be predictive for the bacteriological status of other animals if these animals have some kind of contact with one another. Therefore, it is essential to define the microbiological entity, i.e., a definition of the group of animals for which a sample is predictive. This is a very complicated matter.

Isolators and individually ventilated cages can be defined as two separate microbiological entities, as the idea of the system is to keep the animals out of contact with the surroundings; due to the limited space there is close contact between the animals. Likewise a simple one-room unit used for breeding at a commercial vendor will usually have its own staff and barrier to separate different units from one another; therefore, each of these separate units should be regarded as separate microbiological entities. However, if the unit is separated into different rooms, all of which are served by the same staff and supplied from a common barrier, it is difficult to define the microbiological entity. Some bacteria easily spread from room to to room, while others do not. For example, *P. pneumotropica* is shed by infected animals during their entire lives, is easily spread by passive transport, and readily infects the nasal mucosa of naive animals, while *C. piliforme* has a complicated infective cycle, in which spore-shedding and infection usually are dependent on mother-offspring relationships. So if the unit is used for experiments and not for breeding, there is no mother-offspring relationship and there is no close contact between animals of different experiments. Under such circumstances, it would be safest but also impossible to define each cage as a microbiological entity.

Each experiment consists of several cages, each room contains several experiments, and behind each barrier there may be several rooms. At what level to define the microbiological entity is not a simple matter, but it should be noted that the more subunits to be included, the higher the insecurity of the sampling; for example, if the infection is only found in one subunit, the prevalence is "diluted," which may not be taken into consideration when calculating the sample size.

1.2.3   *Immunosuppression of animals prior to sampling (stress testing)*

Immunosuppression prior to isolation attempts — in order to facilitate the isolation of certain bacteria from healthy animals — is a well-described procedure, often referred to as "stress testing." Table 1.5 shows different treatment regimes for different bacteria. However, nowadays methods of

Table 1.5   Immunosuppression of Animals for Facilitation of the Isolation
of Specified Bacteria and Some Diagnostic Alternatives with Less Impact
on Animal Welfare[a]

Bacteria	Immunosuppressive drug	Dose	Alternative
Clostridium piliforme	Prednisolone	10 mg/kg/day i.p. for 2 days	Serology
Streptobacillus moniliformis	Cyclophosphamide	80 mg/kg i.p.	Cultivation, serology
Citrobacter rodentium, Yersinia spp.	Indomethacin	10 mg/kg oral for 3 days	Cultivation
Salmonella	Chloralhydrate	400 mg/kg i.p.	Cultivation

[a] From Hansen et al.[42] and Juhr.[43]

serology, enriched media, and molecular biology have minimized the need
for using these methods for routine health monitoring. For ethical reasons,
immunosuppression should, therefore, only be used in those cases where no
alternative is available.

1.2.4   Screening vs. profiling

In routine bacteriological monitoring the primary aim is to verify the presence or absence of certain specified bacteria. This is called *screening*. If a
dichotomic identification leads onto a path that does not contain any of those
bacteria screened for, any further identification may be omitted. However,
occasionally there is a need to describe the bacteriological variety or the lack
of such variety. Therefore, all bacterial species are identified as precisely as
possible. This process is called *profiling*.

1.3   Bacteriological examination of diseased animals

Examination of diseased animals is a simpler matter than examination of
healthy animals. The disease symptoms usually derive from certain organs,
an abscess, or the equivalent, which can be sampled directly. Often the
decision to carry out bacteriology is made during a necropsy procedure. It
is, therefore, essential for the bacteriological examination that sterility is
maintained during such a necropsy procedure, until it can be decided
whether bacterial cultivation will be performed or not.

When sampling from diseased organs, the causative agent will often be
found as a pure culture and as such will be much easier to handle. Alternatively, if a pure culture is not found it can often be determined that the
bacterial infection in the organ is secondary to a nonbacterial etiology.

References

1. Hansen, A. K., Health status and the effect of microbial organisms on animal experiments, in *Handbook of Laboratory Animal Science, Vol. I, Animal Selection and Handling in Biomedical Research*, Svendsen, P. and Hau, J., Eds., CRC Press, Boca Raton, FL, 1994, Chap. 11.

2. Lindsey, J. R., Davidson, M. K., Schoeb, T. R., and Cassell, G. H., Murine mycoplasmal infections, diseases, and research complications, in *Complications of Viral and Mycoplasma Infections in Rodents to Toxicology Research*, Ham, T. E., Jr., Ed., Hemisphere Press, Washington D.C., 1985, 91.

3. Broderson, J. R., Lindsey, J. R., and Crawford, J., Role of environmental ammonia in respiratory mycoplasmosis of the rat, *Am. J. Pathol.*, 85, 115, 1976.

4. Schoeb, T. R., Davidson, M. K., and Lindsey, J. R., Intracage ammonia promotes growth of Mycoplasma pulmonis in respiratory tracts of rats, *Infect. Immun.*, 38, 212, 1982.

5. Tvedten, H. W., Whitehair, C. K., and Langham, R. F., Influence of vitamins A and E on gnotobiotic and conventionally maintained rats exposed to Mycoplasma pulmonis, *J. Am. Vet. Med. Assoc.*, 163, 605, 1973.

6. Hoag, W. G., Strout, J., and Meier, H., Epidemiological aspects of the control of Pseudomonas infection in mouse colonies, *Lab. Anim. Care*, 15(3), 217, 1965.

7. Institute of Laboratory Animal Resources, *A Guide to Infectious Diseases of Guinea Pigs, Gerbils, Hamsters and Rabbits*, National Academy of Sciences, Washington, D.C., 1974.

8. Weisbroth, S. H., Bacterial diseases, in *The Laboratory Rat*, Vol. 1, Baker, H. J., Lindsey, J. R., and Weisbroth, S. H., Eds., Academic Press, New York, 1979, 194.

9. Fujiwara, K., Tyzzer's disease, *Jpn. J. Exp. Med.*, 48, 467, 1978.

10. Detmer, A., Hansen, A. K., Dieperink, H., and Svendsen, P., Xylose-positive staphylococci as a cause of respiratory disease in immunosuppressed rats, *Scand. J. Lab. Anim. Sci.*, 18(1), 13, 1990.

11. Dagnæs-Hansen, F. and Bisgaard, M., Biochemical characterization of P. pneumotropica subspp. and their clinical importance in mice, *GV-SOLAS Wissenschaftliche Tagung*, Hannover, 1989.

12. Rygaard, J., *Thymus and Self Immunobiology of the Mouse Mutant Nude*, John Wiley & Sons, London, 1973.

13. Custer, R. P., Outzen, H. C., Eaton, G. J., and Prehn, R. T., Does the absence of immunological surveillance affect the tumour incidence in "nude" mice. First recorded spontaneous lymphoma in a nude mouse, *J. Natl. Cancer. Inst.*, 51, 507, 1973.

14. Cassell, G. H., Derrick Edward Award Lecture. The pathogenic potential of mycoplasmas: Mycoplasma pulmonis as a model, *Rev. Infect. Dis.*, 4(Suppl.), S18-34, 1982.

15. Hansen, A. K., Dagnæs-Hansen, F., and Møllegaard-Hansen, K. E., Correlation between megaloileitis and antibodies to Bacillus piliformis in laboratory rat colonies, *Lab. Anim. Sci.*, 42(5), 449, 1992.

16. Fujiwara, K., Hirano, N., Takenaka, S., and Sato, K., Peroral infection in Tyzzer's disease in mice, *Jpn. J. Exp. Med.*, 43, 33, 1973.

17. Onodera, T. and Fujiwara, K., Nasoencephalopathy in suckling mice inoculated intranasally with the Tyzzer's organism, *Jpn. J. Exp. Med.*, 43, 509, 1973.

18. Jersey, G. C., Whitehair, C. K., and Carter, G. R., Mycoplasma pulmonis as the primary cause of chronic respiratory disease in rats, *J. Am. Vet. Med. Assoc.*, 163, 599, 1972.

19. Hardenbergh, J. G., Epidemic lymphadenitis with formation of abscesses in guinea pigs due to infection with hemolytic streptococci, *J. Lab. Clin. Med.*, 12, 119, 1926.

20. Fortmeyer, H. P., Besondere gesundheitliche Risken der Mutante, in *Thymus-aplastische Maus (nu/nu) Thymusaplastiche Ratte (rnu/rnu) Haltung, Zucht, Versuchsmodelle*, Paul Parey, Berlin, 1981, chap. 3.

21. Trahan, C. J., Stephenson, E. H., Ezzell, J. W., and Mitchell, W. C., Airborne-induced experimental Bordetella bronchiseptica pneumonia in strain 13 guinea pigs, *Lab. Anim.*, 21, 226, 1987.

22. Fujiwara, K., Yamada, A., Ogawa, H., and Oshima, Y., Comparative studies on the Tyzzer's organisms from rats and mice, *Jpn. J. Exp. Med.*, 41(2), 125, 1971.

23. Fujiwara, K., Kurashina, H., Magaribuchi, T., Takenaka, S., and Yokoiyama, S., Further observations on the difference between Tyzzer's organisms from mice and those from rats, *Jpn. J. Exp. Med.*, 43(4), 307, 1973.

24. Hansen, A. K., The use of Mongolian gerbils as sentinels for infection with Bacillus piliformis in laboratory rats, in *Proc. 4th FELASA Symp.*, 449, FELASA, L'Arbresle Cédex, 1992.

25. Van Swieten, M. J., Solleveld, H. A., Lindsey, J. R., deGrott, F. G., Zurcher, C., and Hollander, C. F., Respiratory disease in rats associated with a filamentous bacterium: a preliminary report, *Lab. Anim. Sci.*, 30, 215, 1980.

26. Ganaway, J. R., Spencer, T. H., Moore, T. D., and Allen, A. M., Isolation, propagation and characterization of a newly recognized pathogen, cilia-associated bacillus of rats: an etiological agent of chronic respiratory disease, *Infect. Immun.*, 47, 472, 1983.

27. Brennan, P. C., Fritz, T. E., and Flynn, R. J., The role of Pasteurella pneumotropica and Mycoplasma pulmonis in murine pneumonia, *J. Bacteriol.*, 97, 337, 1969.

28. Moesgaard, F., Nielsen, M. C. L., and Justesen, T., Experimental animal model of surgical wound infection applicable to antibiotic prophylaxis, *Eur. J. Clin. Microbiol.*, 2, 459, 1983.

29. Panton, O. N. M., Smith, J. A., Bell, G. A., Forward, A. D., Murphy, J., and Doyle, P. W., The incidence of wound infection after stapled or sutured bowel anastomosis and stapled or sutured skin closure in humans and guinea pigs, *Surgery*, 98, 20, 1985.

30. Weisbroth, S. H., Bacterial diseases, in *The Laboratory Rat*, Vol. 1, Baker, H. J., Lindsey, J. R., and Weisbroth, S. H., Eds., Academic Press, New York, 1979, 194.

31. Kraft, V., Deeney, A. A., Blanchet, H. M., Boot, R., Hansen, A. K., Hem, A., van Herck, H., Kunstyr, I., Milite, G., Needham, J. R., Nicklas, W., Perrot, A., Rehbinder, C., Richard, Y., and de Vroy, G., Recommendations for the health monitoring of mouse, rat, hamster, guinea pig and rabbit breeding colonies, *Lab. Anim.*, 28, 1, 1994.

32. Rehbinder, C., Baneux, P., Forbes, D., van Herck, H., Nicklas, W., Rugaya, Z., and Winckler, G., Recommendations for the health monitoring of mouse, rat, hamster, guinea pig and rabbit experimental units, *Lab. Anim.*, 30, 193, 1996.

33. Hem, A., Hansen, A. K., Rehbinder, C., Voipio, H. M., and Engh, E., Recommendations for health monitoring of pig, cat, dog and gerbil breeding colonies, *Scand. J. Lab. Anim. Sci.*, 21(3), 97, 1994.
34. Waggie, K., Kagiyama, N., Allen, A. M., and Nomura, T., Manual of Microbiologic Monitoring of Laboratory Animals, NIH Publication No. 94-2498, U.S. Department of Health and Human Services, Public Health Service, National Institutes of Health, National Center for Research Resources, Bethesda, 1994.
35. Hansen, A. K., Andersen, H. V., and Svendsen, O., Studies on the Diagnosis of Tyzzer's Disease in Laboratory Rat Colonies with Antibodies against Bacillus piliformis (Clostridium piliforme), *Lab. Anim. Sci.*, 44(5), 424, 1994.
36. Schwalbe, C. W., Riemann, H. P., and Franti, C. E., *Epidemiology in Veterinary Practice*, Lea & Fibiger, Philadelphia, 1977.
37. Hansen, A. K., The aerobic bacterial flora of laboratory rats from a Danish breeding centre, *Scand. J. Lab. Anim. Sci.*, 19(2), 59, 1992.
38. Hsu, C. K., New, A. E., and Mayo, J. K., Quality assurance of rodent models, in *Animal Quality and Models in Biomedical Research*, Spiegel, A., Erichsen, S., and Solleveld, H.A., Eds., Gustav Fischer Verlag, New York, 1980, 17.
39. Martin, S. W., The evaluation of tests, *Can. J. Comp. Med.*, 41, 19, 1977.
40. Wulff, H. R., *Rational Diagnosis and Treatment*, Blackwell Scientific, Oxford, 1976.
41. Hansen, A. K., Statistical aspects of health monitoring of laboratory animal colonies, *Scand. J. Lab. Anim. Sci.*, 20(1), 11, 1993.
42. Hansen, A. K., Svendsen, O., and Møllegaard-Hansen, K. E., Epidemiological studies of Bacillus piliformis infection and Tyzzer's disease in laboratory rats, *Z. Versuchstierkd.*, 33, 163, 1990.
43. Juhr, N. C., Provocation of latent infections, in *New Developments in Biosciences: Their Implications for Laboratory Animal Science*, Beyen, A. C. and Solleveld, H. A., Eds., Martinus Nijhoff, Dordrecht, 1988, 127.

chapter two

Sampling procedures

Contents

2.1   Planning the work...15
2.2   Euthanasia..15
2.3   Blood sampling 18
2.4   Instruments and sterilizing procedures during sampling....................20
2.5   Opening and inspecting the carcass..21
2.6   Sampling from various organs..21
References...46

2.1   Planning the work

A bacteriological investigation includes several procedures that may be time-consuming, e.g., because the samples at various steps have to be incubated overnight or for a number of days. Therefore, the work has to be carefully planned. For routine sampling this may be done as proposed in Table 2.1. In short, the process of screening for various bacteria in laboratory animals consists of the steps shown in Figure 2.1.

2.2   Euthanasia

If the only procedure to take place is cultivation, the animal will not have to be alive during sampling, so it should humanely be killed prior to the procedure. It is essential that this killing does not interfere with sterility or destroy the organs to be sampled. For small rodents and guinea pigs, the safest way to ensure this is to place the animal in a chamber filled with a mixture of carbon dioxide and oxygen (typically 4:1), which will induce unconsciousness within seconds and death within minutes. If such a chamber is not available rodents may be euthanized by injecting approximately 2 ml/kg body weight of sterile 10% pentobarbitone solution intraperitoneally. This method may interfere with abdominal sampling. Larger animals, such as rabbits, should be given the injection intravenously. Breaking the neck will normally destroy the trachea and thereby prevent proper sampling from this site.

Table 2.1 Example of a Working Plan for Isolation, Cultivation, and Identification of Bacteria from Laboratory Animals

Day	Nonselective media	Selective media
Monday	Sampling and inoculation from the animals	
Tuesday	*First reading of primary plates* Subcultivation of cultivated bacteria	Subcultivation from some selective enrichment broths onto solid media
Wednesday	*Second reading of primary plates* Subcultivation of cultivated bacteria	*Reading and interpreting of solid media* Subcultivation of cultivated bacteria
	Tests for dividing isolated bacteria into major groups or genera	
	Immunological and enzymatic tests	
	Inoculation in test media for identification	
Thursday	*Third reading of primary plates* Subcultivation of cultivated bacteria	Tests for dividing isolated bacteria into major groups or genera
	Tests for dividing isolated bacteria into major groups or genera	Immunological and enzymatic tests
	Immunological and enzymatic tests	Inoculation in test media for identification
	Inoculation in test media for identification	
	Reading test media Inoculating supplementary media	

Friday	Reading supplementary media	Reading test media	Reading test media
Monday	Inoculating supplementary media	Inoculation in test media for identification / Tests for dividing isolated bacteria into major groups or genera. Immunological and enzymatic tests	Inoculating supplementary media
Tuesday	Reading supplementary media	Inoculating supplementary media	Reading supplementary media
Wednesday		Reading supplementary media	

Conclusion and reporting

Note: Often media must be incubated longer than assumed in this table. In addition to the work involved in cultivation, the laboratory may have to do serology and molecular biology directly on samples before a full report can be issued.

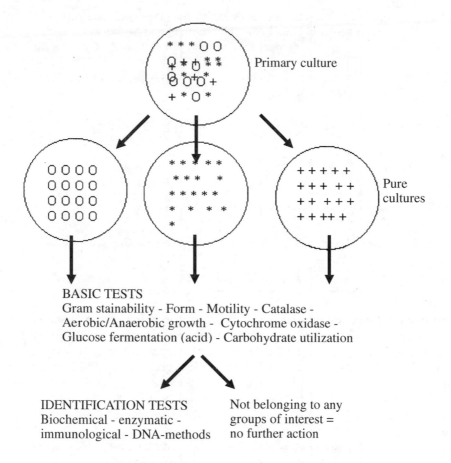

Figure 2.1 The various steps in cultivation and identification of bacteria from laboratory animals. Agar plates are inoculated from a certain organ, which leads to a mixture of several types of bacteria (primary culture), which can be distinguished by their colony morphology. A representative of each type of colony is picked and grown as a pure culture. Each pure culture is subjected to a range of basic tests. If the basic tests allocate the bacteria to groups not containing any of those bacterial species searched for, no further work is done on that culture. If, however, the bacteria is allocated to groups containing bacterial species of interest, a range of tests is performed until the isolate has been identified.

2.3 Blood sampling

If blood has to be sampled for serology, the animal should first be anesthetized and then killed. For most rodents, guinea pigs, and rabbits the preferable anesthetic agent is the combination of fentanyl-fluanisone (Hypnorm®) and midazolam (Dormicum®), which may be used according to Table 2.2. As Hypnorm® is difficult to purchase in the U.S. due to Drug

Table 2.2   Anesthesia of Rodents, Guinea Pigs, and Rabbits
with Fentanyl Fluanisone (Hypnorm®) and Midazolam
(Dormicum®, 5 mg/ml)

Species	Ml mixture/100 g body weight	Injection route
Mouse	0.5–0.75	Subcutaneously
Rat	0.15–0.2	Subcutaneously
Guinea pig	0.6	Intraperitoneally
Rabbit	0.15–0.2	Intramuscularly
	0.04	Slowly intravenously

Note: Both preparations are diluted 1:1 with sterile water, then the two sets of dilutions are mixed 1:1. This mixture is used for anesthesia according to the table. Care should always be taken to inject small rodents subcutaneously to avoid first-pass effect in the portal system, and to inject guinea pigs intraperitoneally as they are stressed by the longer induction time after subcutaneous injection. If the intravenous route is used in the rabbit, it is of the outmost importance that the full dose is given over approx. 5 min or more to avoid hypotensive shock.

Enforcement Agency regulations, U.S. laboratories may have to search for alternatives. Barbiturates, e.g., pentobarbitone in a dose of 50 mg/kg for male rats, 25 mg/kg for female rats, and 45 mg/kg for mice, is frequently used. There is, however, a narrow range between efficient and lethal dose, and hemolysis is often observed in blood samples after barbiturate anesthesia. Another alternative may be to anesthetize the animals with a combination of ketamine and either xylazine or acetylpromazine (Table 2.3).

Table 2.3   Anesthesia of Rodents, Guinea Pigs, and Rabbits with Ketamine
(Vetalar®, Ketaset®, Ketalar®, Ketaminol®), Xylazine (Rompun®, Narcoxyl®),
and Acetylpromazine (Acepromazine®)

Species	Dose			Route
	Ketamine	Xylazine	Acetylpromazine	
Mouse	100	15		Intramuscularly, intraperitoneally
Rat	100	15		Intramuscularly, subcutaneously
Guinea pig	40	5		Intramuscularly, subcutaneously
Hamster	200	10		Intramuscularly, subcutaneously
Gerbil	50	2		Intramuscularly
Rabbit	75		5	Intramuscularly

Data from University of Florida, Animal Resources, Analgesics and Anaesthetic Drug Dosages, http://nersp.nerdc.ufl.edu/~iacuc/drugs.html, 1998.

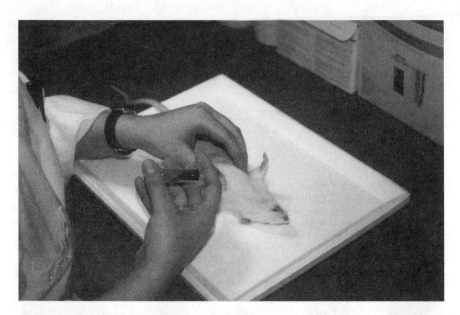

Figure 2.2 Sampling blood from a rat by heart puncture using a vacuum blood sampling glass.

For each animal further information may be found in textbooks on anesthesia, e.g., Flecknell, 1996.[1]

When the animal is sufficiently anesthetized blood is sampled by periorbital puncture, heart puncture (Figure 2.2), or other satisfactory method. Detailed instructions are given by Iwarsson et al., 1994.[2] After blood sampling the animal is killed as described above.

If serum is used instead of plasma it is recommended to place a wooden stick with the blood sample in the tube immediately after sampling. The blood sample is then left at room temperature for at least 3 h, after which most of the coagulum can be removed with the stick (Figure 2.3). The remaining coagulum is removed by centrifugation at 2000 to 3000 rpm for 20 min.

2.4   Instruments and sterilizing procedures during sampling

Sampling is done with a platin needle of 0.5 mm in size, for example. The needle is flame-sterilized between each sampling. To open the animal and the various organs it is necessary to have at least a scalpel, a pair of scissors, and a pair of tweezers, which must be sterile before every sampling. Sterilization is achieved by cleaning the instruments in a glass of ethanol (70 or 93%) and holding them in a flame between each sampling. The burning instruments are then placed upon, e.g., a glass rack (Figure 2.4), until the fire has ceased.

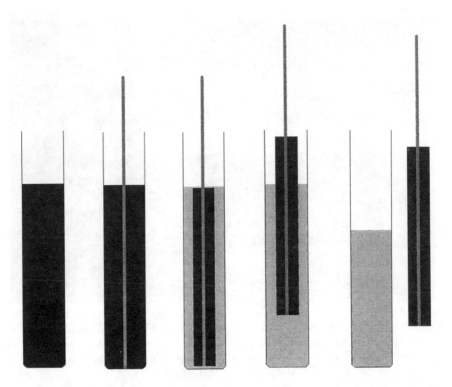

Figure 2.3 Removal of the coagulum from a blood sample might be facilitated by placing a wooden stick with the blood sample in the tube immediately after sampling. The blood sample is then left at room temperature for at least 3 h, after which most of the coagulum can be removed with the stick.

2.5 Opening and inspecting the carcass

The animal is placed on its back on a polystyrene tray and its four feet are attached to the tray with pins (Figure 2.5). The entire thoracic and abdominal wall may be disinfected with 70% ethanol, which should be allowed to evaporate before any procedures are performed. This disinfection procedure is not 100% necessary for sterile sampling and may be omitted.

2.6 Sampling from various organs (Figures 2.6 to 2.35)

If the reason for examination is some kind of disease in the animal, which organ to sample may be rather straightforward. If a healthy animal is sampled for investigations for one or several bacteria, the choice of organs to be sampled must be based on scientific judgment. In Tables 2.4 (also, Figures 2.6 to 2.17) and 2.5 (includes Figures 2.25 to 2.35) information is given on which organs to sample in the search for certain bacteria. Further descriptions are given for each agent in the specific chapters. Programs

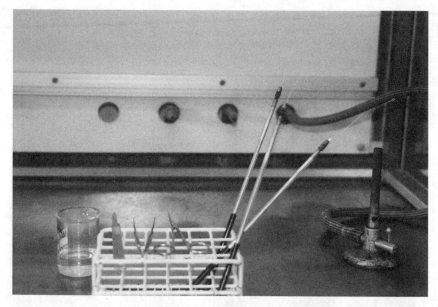

Figure 2.4 Instruments for bacteriological sampling. The instruments are dipped into 70% ethanol in the little glass, and while still wet from the alcohol they are held into the burning flame and placed upon the glass rack until the alcohol has ceased burning. Care should be taken to avoid touching animals disinfected with ethanol with the instruments while the alcohol is still burning.

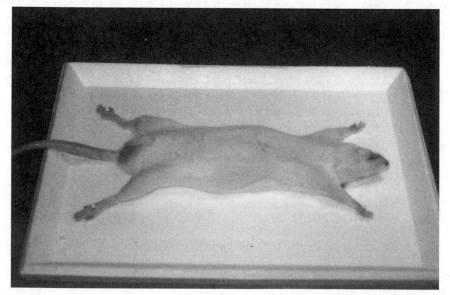

Figure 2.5 The animal is placed on its back on a polystyrene tray with its four feet attached to the tray by pins.

Table 2.4 Bacteriological Sampling Sites Accessible from the Body Surface and Bacterial Species Likely To Be Isolated from These Sites

	Healthy animals	Diseased animals	How to sample
Skin	*Staphylococcus, Pseudomonas* spp.	*Corynebacterium kutscheri, Staphylococcus, Pseudomonas* spp.	The area is scraped with a scalpel and the scraping is transferred to the medium (Figure 2.6); on an agar plate the material is streaked with the platin needle (Figure 2.7)
Nose	*Bordetella bronchiseptica, Mycoplasma* spp., *Pasteurellaceae, Pseudomonas* spp., *Streptobacillus moniliformis*, β-hemolytic streptococci, *Streptococcus pneumoniae*	*Bordetella bronchiseptica, Klebsiella pneumoniae, Mycoplasma* spp., *Pasteurellaceae, Pseudomonas* spp., *Streptobacillus moniliformis*, β-hemolytic streptococci, *Streptococcus pneumoniae*	The skin around the nostrils is removed with scissors (Figure 2.8); the platin needle is introduced deeply into both sides of the nasal cavity (Figure 2.9)
Conjunctiva	β-Hemolytic streptococci	*Pasteurellaceae, Staphylococcus*, β-hemolytic streptococci	The conjunctiva is touched with the platin needle (Figure 2.10)
Middle ear	*Mycoplasma* spp.	*Corynebacterium kutscheri, Mycoplasma* spp., *Pasteurellaceae, Pseudomonas* spp., *Staphylococcus*, β-hemolytic streptococci, *Streptococcus pneumoniae*	The entire auricle is cut off with scissors (Figure 2.11); the accessible membrane (Figure 2.12) is opened with a scalpel, and the platin needle is introduced into the middle ear (Figure 2.13)

continued

Table 2.4 (continued)   Bacteriological Sampling Sites Accessible from the Body Surface and Bacterial Species Likely To Be Isolated from These Sites

	Healthy animals	Diseased animals	How to sample
Trachea	*Bordetella bronchiseptica, Mycoplasma* spp., Pasteurellaceae, *Pseudomonas* spp., *Streptobacillus moniliformis*, β-hemolytic streptococci, *Streptococcus pneumoniae*	*Bordetella bronchiseptica, Klebsiella pneumoniae, Mycoplasma* spp., Pasteurellaceae, *Pseudomonas* spp., *Streptobacillus moniliformis*, β-hemolytic streptococci, *Streptococcus pneumoniae*	The animal is placed on its back, the skin in the ventral neck region is opened, and the muscles over the trachea are bluntly spread (Figure 2.14); the trachea is lifted by placing the scissors underneath it (Figure 2.15), and the trachea is opened with a longitudinal cut (Figure 2.16); the needle is introduced in both the cranial and the caudal direction (Figure 2.17)
Feces	*Campylobacter* spp., *Citrobacter rodentium, Mycobacterium* spp., *Pseudomonas* spp., Salmonellae, *Yersinia pseudotuberculosis*	*Campylobacter* spp., *Citrobacter rodentium, Pseudomonas* spp., Salmonellae, *Yersinia pseudotuberculosis*	A fecal pellet is dropped into a broth or suspended in 1 ml sterile saline; with the platin needle the suspension is streaked upon an agar plate

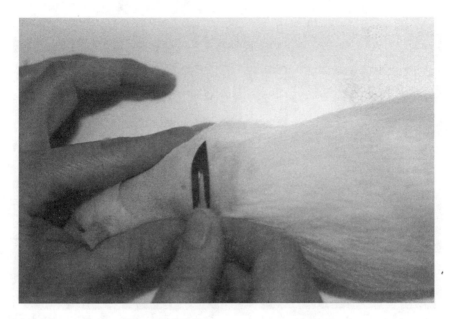

Figure 2.6 An area of the skin to be sampled is scraped with a scalpel and the scraping is transferred to the medium.

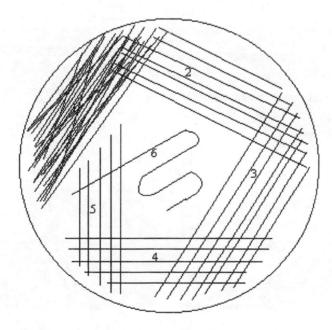

Figure 2.7 Correct streaking of an agar plate.

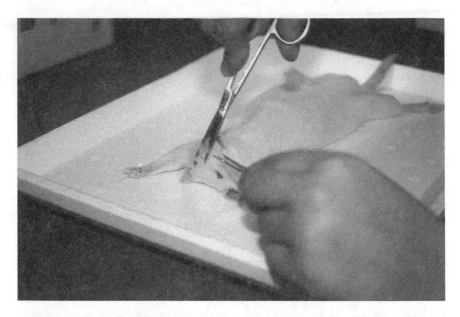

Figure 2.8   The skin around the nostrils is removed with scissors to facilitate
sampling from the nose.

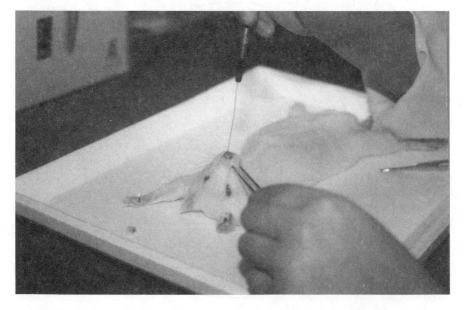

Figure 2.9   The platin needle is introduced deeply into both sides of the nasal cavity.

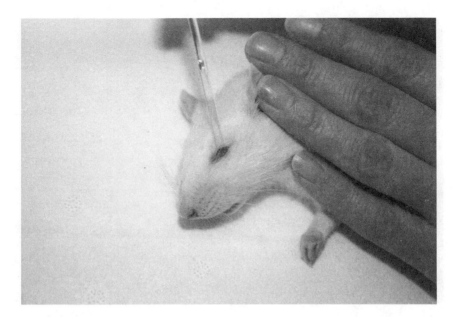

Figure 2.10 The conjunctiva is touched with the needle (in this case a disposable needle was used).

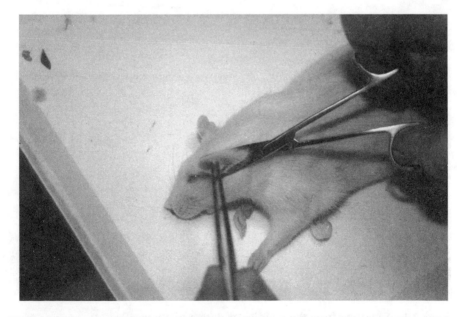

Figure 2.11 Sampling from the middle ear is facilitated by removing the entire auricle.

Figure 2.12 After removing the auricle the membrane is accessible and may be opened with a scalpel.

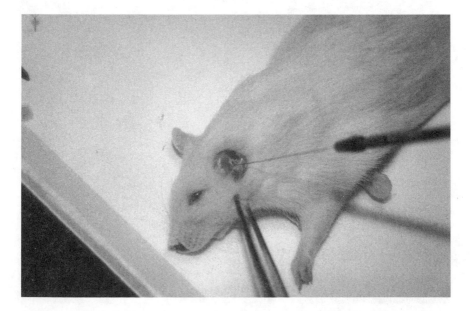

Figure 2.13 Sampling from the middle ear with a platin needle.

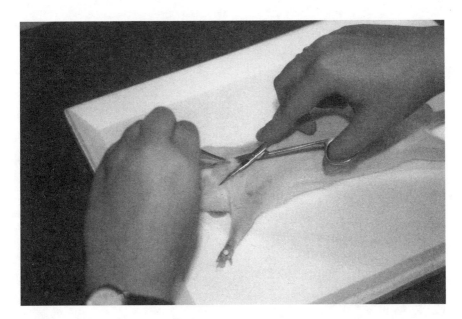

Figure 2.14 The skin in the ventral neck region is opened to access the trachea for sampling, and the muscles over the trachea are bluntly spread.

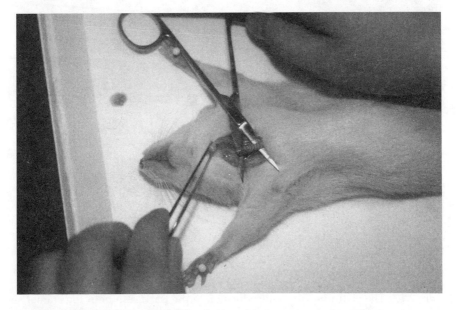

Figure 2.15 The trachea is lifted by placing the tweezers underneath it.

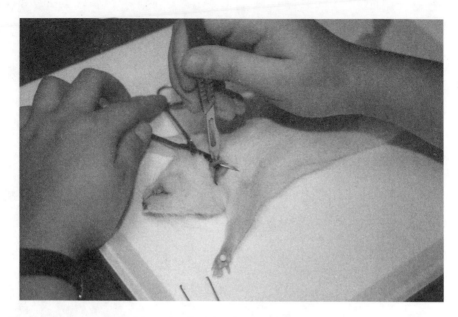

Figure 2.16 The trachea is opened with a longitudinal cut.

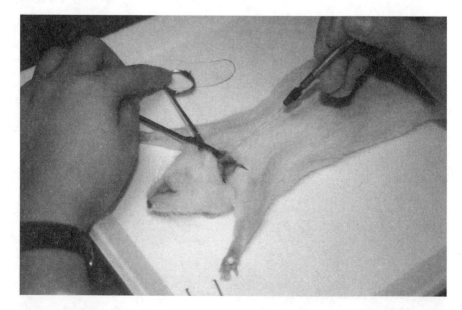

Figure 2.17 The platin needle is introduced into the trachea in both the cranial and the caudal direction.

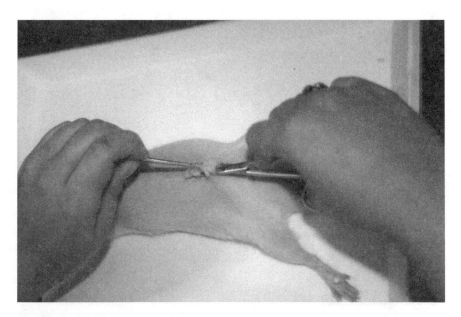

Figure 2.18 The skin is cut to access the abdomen.

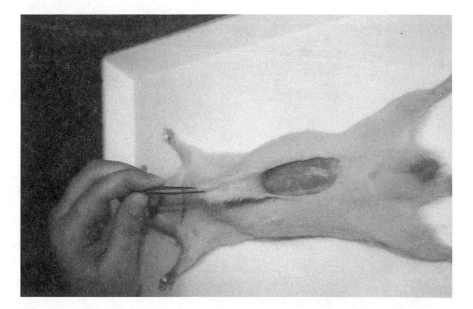

Figure 2.19 The abdominal skin is removed in the longitudinal direction of the animal all the way from above the pecten ossis pubis to the sternal manubrium.

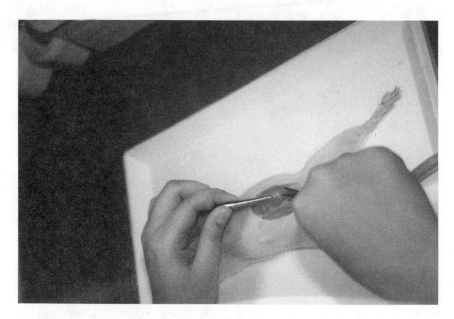

Figure 2.20 A piece of the abdominal wall with muscles and peritoneum is removed from the pecten ossis pubis to the manubrium to access the abdomen.

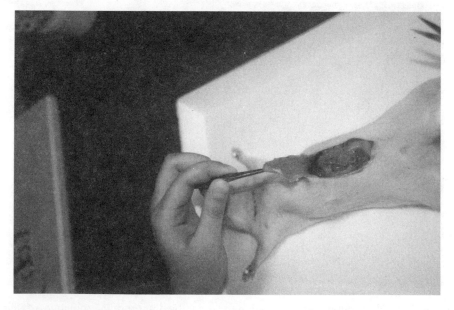

Figure 2.21 After removing the abdominal skin, muscles, and peritoneum the abdominal organs can be located and sampled.

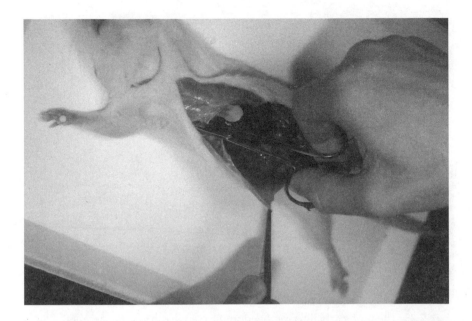

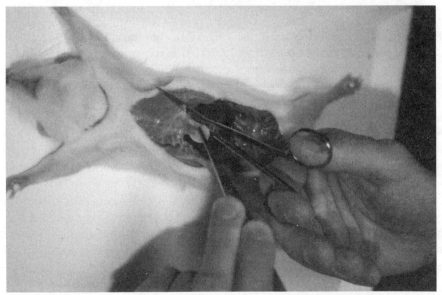

Figure 2.22 (a and b) After removing the ventral skin, the thorax is opened by lifting the manubrium sternum with tweezers and making a small hole into the diaphragm with scissors. The thoracic wall is opened with a cut in both sides directed from this opening toward the forelimbs.

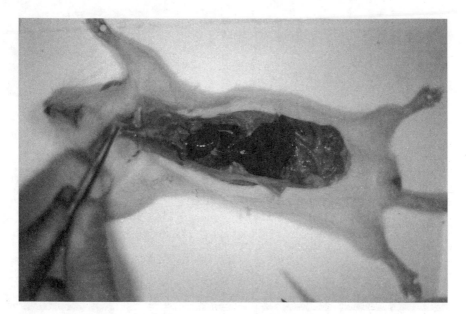

Figure 2.23 The thoracic wall is removed in the cranial direction to access the thorax for sampling.

should be planned individually according to the aims. For routine monitoring of rodents, guinea pigs, and rabbits it is recommended to include at least the nose, trachea, cecum, and the genitals in the examination. Examples of programs for different species are shown in Tables 2.6 to 2.9.

The abdomen is opened by first removing a piece of skin with scissors in the longitudinal direction of the animal all the way from above the pecten ossis pubis to the sternal manubrium (Figures 2.18 and 2.19). A piece of the abdominal wall with muscles and peritoneum is removed from the pecten ossis pubis to the manubrium (Figure 2.20). The abdominal organs then can be located and sampled (Figure 2.21). To access the thorax the skin above the sternum is removed. The thorax is opened by lifting the manubrium sternum with tweezers and with scissors making a small hole into the diaphragm. A cut in both sides directed from this opening toward the forelimbs opens the thoracic wall (Figures 2.22a,b) and moves it in the cranial direction (Figure 2.23).

Table 2.5 Bacteriological Sampling Sites Accessible Only after Opening Thorax or Abdomen, and Bacterial Species Likely To Be Isolated from These Sites

	Healthy animals	Diseased animals	How to sample
Lungs	*Mycoplasma* spp.	*Bordetella bronchiseptica, Corynebacterium kutscheri, Corynebacterium pneumoniae, Mycobacterium* spp., *Mycoplasma* spp., *Pasteurellaceae, Pseudomonas* spp., β-hemolytic streptococci, *Streptococcus pneumoniae*	A piece of each of the lungs is cut off with scissors and the cut surface is pressed into the surface of an agar plate (Figure 2.24); the material is streaked with the platin needle (Figure 2.7); for propagation the piece of lung is simply dropped into an enrichment broth
Cecum	*Campylobacter* spp., *Citrobacter rodentium, Escherichia coli, Klebsiella pneumoniae, Pasteurellaceae, Salmonella, Yersinia pseudotuberculosis, Helicobacter* spp.	*Campylobacter* spp., *Citrobacter rodentium, Escherichia coli, Klebsiella pneumoniae, Pasteurellaceae, Salmonella, Yersinia pseudotuberculosis, Helicobacter* spp.	A piece of the cecal wall (major curve, upper side, midsection) is cut off with scissors (Figure 2.25) and the platin needle is dipped into the cecal contents (Figure 2.26), and then either streaked upon an agar plate or dipped into a broth
Liver	*Helicobacter* spp.	*Corynebacterium kutscheri, Escherichia coli, Helicobacter* spp., *Pasteurellaceae*	A lobe of the liver is grasped with the tweezers and the grasped piece is cut off (Figure 2.31); the piece is plunged into a broth or the cut surface is touched upon the agar plate (Figure 2.32), which then is streaked with the platin needle *continued*

Table 2.5 (continued) Bacteriological Sampling Sites Accessible Only after Opening Thorax or Abdomen, and Bacterial Species
Likely To Be Isolated from These Sites

	Healthy animals	Diseased animals	How to sample
Genitals	*Corynebacterium kutscheri, Escherichia coli, Mycoplasma spp., Pasteurellaceae, β-hemolytic streptococci, Streptococcus pneumoniae, Listeria spp.*	*Corynebacterium kutscheri, Escherichia coli, Mycoplasma spp., Pasteurellaceae, β-hemolytic streptococci, Streptobacillus moniliformis, Streptococcus pneumoniae, Listeria spp.*	♀: The vulvae are cut off with scissors (Figure 2.27) and the platin needle is introduced deeply into the vagina (Figure 2.28) ♂: The preputium is cut off (Figure 2.29) and the glans is touched with the platin needle (Figure 2.30); the needle then is either streaked upon an agar plate or dipped into a broth
Spleen		*Pasteurellaceae, Streptobacillus moniliformis, Streptococcus pneumoniae*	As for the liver (see above)
Kidneys		*Corynebacterium kutscheri, Escherichia coli*	The kidney is halved in the pole-to-pole direction (Figure 2.33) and the pelvic surface is touched with the platin needle (Figure 2.34)

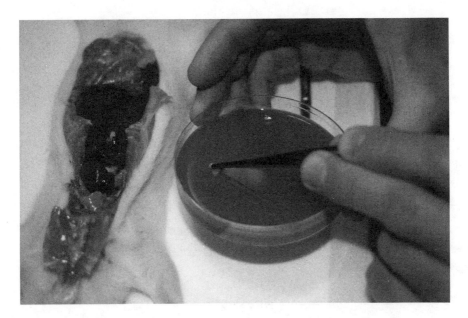

Figure 2.24 The cut surface of the lung piece is pressed onto the surface of an agar plate.

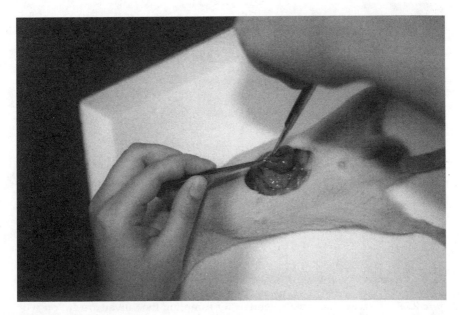

Figure 2.25 A piece of the cecal wall (major curve, upper side, midsection) is cut off with the scissors to access the cecal contents for sampling.

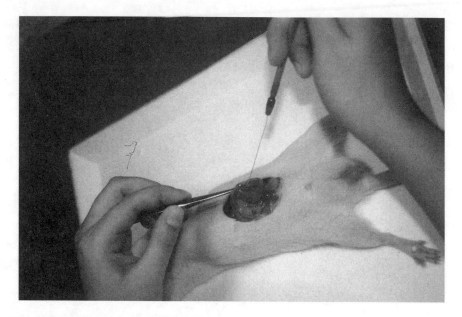

Figure 2.26 Sampling from the cecal contents.

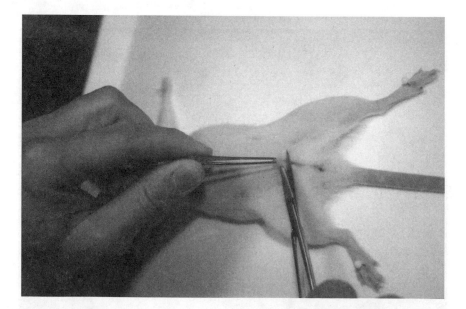

Figure 2.27 To access the female genitals for sampling the vulvae are cut off with the scissors.

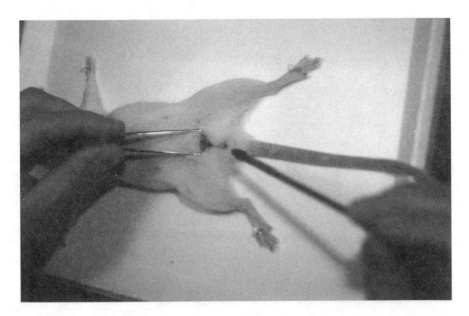

Figure 2.28 The platin needle is introduced deeply into the vagina after the vulvae have been removed.

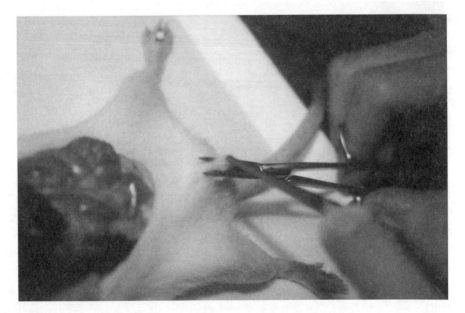

Figure 2.29 To access the male genitals for sampling the preputium is cut off.

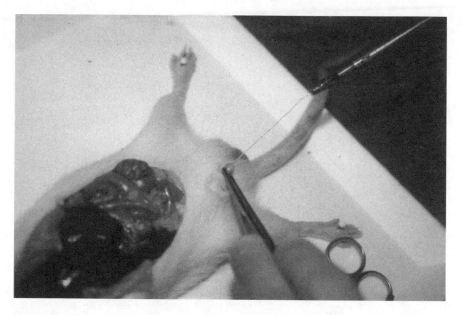

Figure 2.30 The glans is touched with the platin needle after the preputium has been removed.

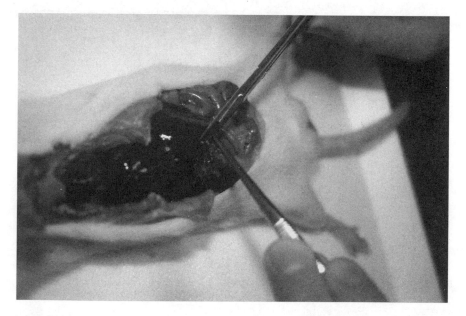

Figure 2.31 A lobe of the liver is grasped with the tweezers and the grasped piece is cut off.

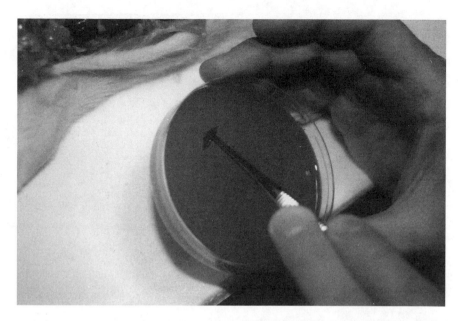

Figure 2.32 The cut surface of a piece of the liver is streaked on an agar plate.

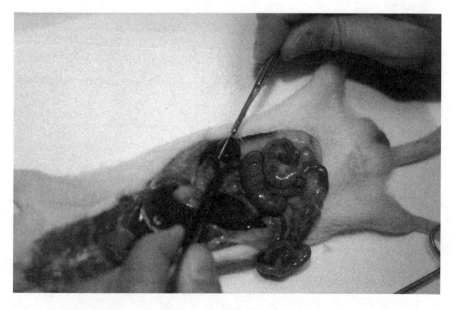

Figure 2.33 The kidney is halved in the pole-to-pole direction to access the renal pelvis for sampling.

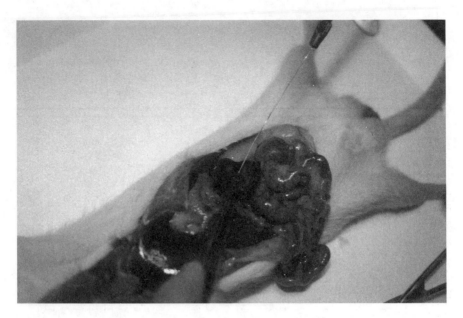

Figure 2.34 The surface of the renal pelvis is touched with the platin needle.

Table 2.6 Example of a Program for Routine Bacteriological Examination of Laboratory Mice and Rats

Organ	Medium	Incubation	Bacteria screened for
Nose	Chocolate agar	Microaerophilic, 37°C, 24 h	*Bordetella bronchiseptica, Haemophilus* spp., *Klebsiella pneumoniae, Pasteurella* spp., *Streptobacillus moniliformis,* β-hemolytic streptococci, *Streptococcus pneumoniae*
Trachea	20% serum agar	Microaerophilic, 37°C, 3 days	*Streptobacillus moniliformis*
	Chocolate agar	Microaerophilic, 37°C, 24 h	*Bordetella bronchiseptica, Haemophilus* spp., *Klebsiella pneumoniae, Pasteurella* spp., *Streptobacillus moniliformis,* β-hemolytic streptococci, *Streptococcus pneumoniae*
	20% serum agar	Microaerophilic, 37°C, 3 days	*Streptobacillus moniliformis*
Cecum	Blood agar with chloral hydrate	Aerobic, 37°C, 24 h	*Citrobacter rodentium* (mice only), *Klebsiella pneumoniae, Pasteurella* spp., *Salmonella* spp., *Yersinia* spp.
	Selenite broth	Aerobic, 37°C, 24 h followed by streaking on BPLS agar	*Salmonella* spp., *Yersinia* spp.
Genitals	TVP agar	Microaerophilic, 37°C, 5 days	*Campylobacter* spp., *Helicobacter* spp.
	Chocolate agar	Microaerophilic, 37°C, 24 h	*Bordetella bronchiseptica, Citrobacter rodentium* (mice only), *Haemophilus* spp., *Klebsiella pneumoniae, Pasteurella* spp., *Streptobacillus moniliformis,* β-hemolytic streptococci, *Streptococcus pneumoniae*
Serum (antibodies)			CAR bacillus (rats only) — ELISA
			Clostridium piliforme — ELISA
			Helicobacter hepaticus (mice only) — ELISA
			Leptospira spp. — Agglutination
			Mycoplasma pulmonis — ELISA
			Pasteurellaceae — ELISA

Table 2.7 Example of a Program for Routine Bacteriological Examination of Laboratory Hamsters and Gerbils

Organ	Medium	Incubation	Bacteria screened for
Nose	Chocolate agar	Microaerophilic, 37°C, 24 h	Bordetella bronchiseptica, Klebsiella pneumoniae, Pasteurella spp., β-hemolytic streptococci, Streptococcus pneumoniae
Trachea	Chocolate agar	Microaerophilic, 37°C, 24 h	Bordetella bronchiseptica, Klebsiella pneumoniae, Pasteurella spp., β-hemolytic streptococci, Streptococcus pneumoniae
Cecum	Blood agar with chloral hydrate	Aerobic, 37°C, 24 h	Klebsiella pneumoniae, Pasteurella spp., Salmonella spp., Yersinia spp.
	Selenite broth	Aerobic, 37°C, 24 h followed by streaking on BPLS agar	Salmonella spp., Yersinia spp.
Genitals	Chocolate agar	Microaerophilic, 37°C, 24 h	Bordetella bronchiseptica, Klebsiella pneumoniae, Pasteurella spp., β-hemolytic streptococci, Streptococcus pneumoniae
Serum (antibodies)		ELISA	Clostridium piliforme

Table 2.8 Example of a Program for Routine Bacteriological Examination of Laboratory Guinea Pigs

Organ	Medium	Incubation	Bacteria screened for
Nose	Chocolate agar	Microaerophilic, 37°C, 24 h	Bordetella bronchiseptica, Klebsiella pneumoniae, Pasteurella spp., Streptobacillus moniliformis, β-hemolytic streptococci, Streptococcus pneumoniae
	20% serum agar	Microaerophilic, 37°C, 3 days	Streptobacillus moniliformis
Trachea	Chocolate agar	Microaerophilic, 37°C, 24 h	Bordetella bronchiseptica, Klebsiella pneumoniae, Pasteurella spp., Streptobacillus moniliformis, β-hemolytic streptococci, Streptococcus pneumoniae
	20% serum agar	Microaerophilic, 37°C, 3 days	Streptobacillus moniliformis

Organ	Medium	Incubation	Bacteria screened for
Cecum	Blood agar with chloral hydrate Selenite broth	Aerobic, 37°C, 24 h Aerobic, 37°C, 24 h followed by streaking on BPLS agar	*Klebsiella pneumoniae, Pasteurella* spp., *Salmonella* spp., *Yersinia* spp. *Salmonella* spp., *Yersinia* spp.
Genitals	Chocolate agar	Microaerophilic, 37°C, 24 h	*Bordetella bronchiseptica, Klebsiella pneumoniae, Pasteurella* spp., *Streptobacillus moniliformis,* β-hemolytic streptococci, *Streptococcus pneumoniae*
Serum (antibodies)			*Clostridium piliforme* ELISA

Table 2.9 Example of a Program for Routine Bacteriological Examination of Laboratory Rabbits

Organ	Medium	Incubation	Bacteria screened for
Nose	Chocolate agar	Microaerophilic, 37°C, 24 h	*Bordetella bronchiseptica, Klebsiella pneumoniae, Pasteurella* spp., β-hemolytic streptococci, *Streptococcus pneumoniae*
Trachea	Chocolate agar	Microaerophilic, 37°C, 24 h	*Bordetella bronchiseptica, Klebsiella pneumoniae, Pasteurella* spp., β-hemolytic streptococci, *Streptococcus pneumoniae*
Cecum	Blood agar with chloral hydrate Selenite broth	Aerobic, 37°C, 24 h Aerobic, 37°C, 24 h followed by streaking on BPLS agar	*Klebsiella pneumoniae, Pasteurella* spp., *Salmonella* spp., *Yersinia* spp. *Salmonella* spp., *Yersinia* spp.
Genitals	Chocolate agar	Microaerophilic, 37°C, 24 h	*Bordetella bronchiseptica, Klebsiella pneumoniae, Pasteurella* spp., *Streptobacillus moniliformis,* β-hemolytic streptococci, *Streptococcus pneumoniae*
Serum (antibodies)			*Clostridium piliforme* ELISA *Treponema cuniculi* IFA

References

1. Flecknell, P.A., *Laboratory Animal Anaesthesia*, Academic Press, London, 1996.
2. Iwarsson, K., Lindberg, L., and Waller, T., Common non-surgical techniques, in *A Handbook of Laboratory Animal Science, Vol. 1. Animal Selection and Handling in Biomedical Research*, Svendsen, P., and Hau, J., Eds., CRC Press, Boca Raton, FL, 1994, chap. 16.
3. University of Florida, Animal Resources, Analgesics and Anaesthetic Drug Dosages, http://nersp.nerdc.ufl.edu/~iacuc/drugs.html, 1998.

chapter three

Cultivation methods

Contents

3.1 The choice of media 47
3.2 Incubation of media 49
3.3 Isolation of bacteria 51
References 54

3.1 The choice of media

Most screening programs consist of a combination of selective and non-selective methods. If the aim of the investigation is *profiling*, nonselective media are a must in order to support as many bacteria as possible. Also, when cultivating from pathologically changed organs, the nonselective medium is necessary for diagnosing nonspecific bacterial infections, which are often secondary to some other cause of disease.

The most common nonselective medium is 5% blood agar (Table 3.1). This medium will support the growth of both bacteria with low nutrient demands, e.g., Enterobacteriaceae or Micrococcaceae, as well as bacteria that will only grow in enriched media, e.g., Streptococcaceae. An advantage of this medium is that hemolysis is directly observable on the primary culture. However, this medium may be insufficient for some groups of bacteria, such as Pasteurellaceae, which is an extremely important group in laboratory animal bacteriology. For example, primary growth of *Pasteurella pneumotropica* is only supported by some types of blood, a phenomenon which is not clearly understood, i.e., it is difficult to give guidance for which types of blood to be used. It is therefore necessary to test blood from different sources in each laboratory before the method is reliable. Furthermore, blood agar does not support the growth of *Haemophilus* spp. Instead, chocolate agar (Table 3.2) is likely to support growth of most *Pasteurella* spp. This agar is also suitable for anaerobic cultivation if 0.001 g

Table 3.1 Example of a Recipe for 5% Blood Agar[a]

Sterile, deionized water	1000 ml
Magnesium sulfate, $7H_2O$	0.1 g
Manganese chloride, $7H_2O$	0.0067 g
Disodium hydrogenephosphate, $12H_2O$	0.0067 g
Caseine hydrolysate	5.0 g
Yeast extract	3.0 g
Potassium chloride	6.67 g
Sebacic acid	0.01 g
Agar	10.0 g
Peptone	5.0 g
Defibrinized horse blood	50 ml
Cysteine HCL	0.05 g
Sodium pyruvate	2.0 g

Note: pH is stabilized at 7.4.

[a] An agar base, which only has to be supplied with blood, is available from most producers of bacteriological media and may be used as an alternative.

Table 3.2 Example of a Recipe for Chocolate Agar[a]

Sterile, deionized water	1000 ml
Peptone	15.0 g
Yeast extract	1.0 g
Sodium chloride	5.0 g
Citric acid	0.15 g
Dipotassium hydrogenephosphate	0.0067
Starch	2.0 g
Agar	10.0 g
NAD	0.0004 g
Autolysed liver	20.0 ml
Glucose	1.5 g
Defibrinized horse blood	70 ml[b]

Note: pH is stabilized at 7.2.

[a] If the agar is to be used for anaerobic cultivation, 0.001 g of vitamin K and 0.55 g of cysteine HCL per 1000 ml are added to support growth of *Bacteroides* spp.

[b] Briefly heat the blood shortly before use.

Data from Gibbons, R.J. and MacDonald, J.B., *J. Bacteriol.*, 80, 164, 1960; and Møller, V. and Reyn, A., *Bull. WHO*, 32, 471, 1965.

of vitamin K and 0.55 g of cysteine HCL per 1000 ml are added to support growth of *Bacteroides* spp.

A main problem when using nonselective media is swarming of *Proteus* spp., a phenomenon that may totally inhibit the isolation of anything else. This is especially observed after cultivation from the cecum and the genitals. This is avoided by the addition of either detergents, antibiotics, antisera, or anesthetics to the medium. Detergents and anesthetics are preferable as they do not hinder the isolation of *Proteus* itself, only the swarming. Dodecyle-benzole sulfonate (0.005%) or chloral hydrate (0.1%) is usable.

For many bacteria efficient selective, indicative, or combined selective-indicative media are optimal.Typically, this includes primary inoculation of a selective enrichment broth for propagation followed by the subsequent streaking of an indicative agar. This is especially usable for bacteria only found in low numbers in each animal. Such media are described in the specific chapters of this book.

3.2   Incubation of media

Nonselective media are generally incubated at 37°C. Depending on which bacteria are included in a screening program incubation may be either aer-obic, microaerophilic, or anaerobic. For aerobic incubation the media are simply placed in the incubator. Microaerophilic incubation may be achieved either in a carbon dioxide incubator or by certain commercial systems for microaerophilic incubation. Such systems usually consist of a package with a chemical that can be activated by the addition of water. An example is the Anaerocult C® from Merck, Germany. The agar plates and the microaero-philic generation system are placed together in either a sealed plastic pack (Figure 3.1) or a closed jar (Figure 3.2). In the same way an anaerobic culti-vation environment may be created using systems that will generate an anaerobic atmosphere, e.g., the Anaerocult A® from Merck (Germany). In the jar or plastic bag an indicator stick, also available from the same commercial suppliers, should be placed. A simple reagent for anaerobic incubation may be produced on site in the laboratory (Table 3.3).

Table 3.3 A Simple Reagent Usable for Creating an Anaerobic Environment in a Jar or Sealed Plastic Bag

Pyrogallol	50 g
Potassium carbonate	50 g
Terra silica	250 g

Note: The compounds are efficiently mixed together. Can be maintained in packages of approx. 2 g in an exicator for about 1 month.

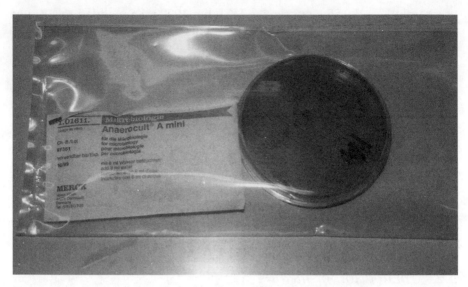

Figure 3.1 An agar plate packed with a microaerophilic generation system in a sealed plastic bag.

Figure 3.2 An anaerobic jar and an anaerobic atmosphere generation kit. The kit is flushed with water and placed with the agar plates in the closed jar.

Aerobic or microaerophilic nonselective media are normally incubated for 18 to 24 h, inspected, reincubated for another 18 to 24 h, reinspected, and reincubated for the last 18 to 24 h, inspected and discarded. Anaerobic media are generally inubated for 72 h before opening the incubation system.

Selective or indicative systems for specific bacteria should be incubated as described in the more specific chapters of this book.

3.3   Isolation of bacteria

Only pure, noncontaminated cultures should be subjected to identification. To make a pure culture from every single colony from the primary plates would be virtually impossible. Therefore, a representative of each morphologically distinct type on the plate is picked up. Often, to make the work more rational, plates from different organs, different animals, or both are grouped, and only one representative of each morphologically distinct colony type within each group is picked up and grown as a pure culture. To do this in a scientifically valid way, the group within which colonies are compared must be clearly defined (Table 3.4), e.g., as either within the same animal but from different organs, or within the same organs but from different animals.

Table 3.4   Isolation of Bacteria from Primary Plates

Organ	\multicolumn{10}{c}{Animal number}

Organ	1	2	3	4	5	6	7	8	9	10
\multicolumn{11}{l}{**A. Individual Animal/Individual Organ Isolation**}
Nose	1, 2, 3, 4	5, 6, 7	8, 9, 10, 11	12, 13	14, 15, 16	17	18, 19	20, 21, 22, 23, 24, 25	26, 27, 28,	29, 30
Trachea	31		32	33, 34	35, 36	37, 38, 39	40			41
Cecum	42, 43, 44, 45	46, 47, 48	49, 50, 51, 52	53, 54, 55, 56, 57	58, 59, 60, 61	62, 63, 64, 65, 66, 67	68, 69, 70	71, 72, 73, 74, 75	76, 77, 78, 79, 80	81, 82, 83, 84
Genitalia	85, 86	87, 88, 89	90, 91, 92	93, 94, 95, 96	97, 98	99, 100	101, 102, 103	104, 105, 106, 107, 108	109, 110, 111, 112, 113	114, 115, 116, 117, 118

continued

Table 3.4 (continued) Isolation of Bacteria from Primary Plates

	Animal number									
Organ	1	2	3	4	5	6	7	8	9	10
B. Individual Animal/Grouped Organ Isolation										
Nose	1,2, 3,4	9, 10, 11	17, 18, 19, 20	27, 28	36, 37, 38	44	56, 57	66, 67, 68, 69, 70, 71	80, 81, 82	92, 93
Trachea	2		18	27, 29	36, 38	45, 46, 47	58			94
Cecum	5,6, 7,8	12, 13, 14	21, 22, 23, 24	30, 31, 32, 33, 34	39, 40, 41, 42	48, 49, 50, 51, 52, 53	59, 60, 61	72, 73, 74, 75, 76	83, 84, 85, 86, 87	95, 96, 97, 98
Genitalia	7,1	13, 15, 16	17, 25, 26	31, 33, 34, 35	41, 43	54, 55	56, 60, 62	66, 75, 77, 78, 79	86, 88, 89, 90, 91	96, 99, 100, 101, 102
C. Grouped Animal/Individual Organ Isolation										
Nose	1,2, 3,4	2,4, 5	1,2, 5,6	12, 13	2,4, 5	1	1,3	1,3, 4,5, 7,8	3,4, 8	2,6
Trachea	9		9	9, 10	9, 10	9, 10, 11	12			10
Cecum	13, 14, 15, 16	13, 14, 15	13, 14, 15, 16	13, 14, 15, 16, 17	13, 14, 16, 17	13, 14, 15, 16, 17, 18	15, 17, 18	14, 15, 16, 17, 18	14, 15, 16, 17, 18	15, 17, 18, 19
Genitalia	20, 21	20, 21, 22	20, 21, 22	20, 22, 23, 24	23, 24	25, 26	20, 22, 23	20, 21, 22, 24, 26	20, 21, 22, 24, 26	20, 21, 22, 23, 27

Table 3.4 (continued) Isolation of Bacteria from Primary Plates

Organ	Animal number									
	1	2	3	4	5	6	7	8	9	10

D. Grouped Animal/Grouped Organ Isolation

Organ	1	2	3	4	5	6	7	8	9	10
Nose	**1, 2, 3, 4**	2, 4, 5	1, 2, 5, 6	2, 4	2, 4, 5	1	1, 3	1, 3, 4, 5, 7, 8	3, 4, 8	2, 6
Trachea	2		2	2, 5	2, 5	2, 5, 6	7			5
Cecum	**8, 9, 10, 11**	8, 9, 10	8, 9, 10, 11	8, 9, 10, 11, **12**	8, 9, 11, 12	8, 9, 10, 11, 12, 13	10, 12, 13	9, 10, 11, 12, 13	9, 10, 11, 12, 13	10, 12, 13, **14**
Genitalia	1, **15**	1, 15, 12	1, 15, 12	1, 12, 11, **16**	11, 16	14, **17**	1, 12, 11	1, 15, 12, 16, 17	1, 15, 12, 16, 17	1, 15, 12, 11, **18**

Note: In this example ten animals were sampled, and inoculations were made from the nose, the trachea, the cecum, and the genitals from each animal on, e.g., chocolate agar. After incubation for 18 to 24 h each plate contains a number of morphologically distinct bacterial colonies. The four report sheets (A to D) in this table show four different systems for isolation of bacteria from the primary plates. Each number represents one specific type of colony, but the colony is only picked up from that plate in which it has been given in bold in the schedule, i.e., the same number given at different locations in the schedule means that the two colonies have been regarded as morphologically alike. If the bacteria of each plate are only compared with bacteria on the same plate and not with any bacteria found on the other plates, i.e., each organ of each animal is examined independently of all other organs whether from the same or from the other animals (A), 118 pure cultures have to be made and afterward each must be passed through an identification process. This is a very safe system, but also extremely expensive and time-consuming. If, instead, the agar plates of the same animal are compared with one another, the number of isolates can be reduced to 102 (B). Alternatively, the individual animals may be kept separate, and the agar plates from the same organs may be compared with one another, which in this example reduces the number of isolates to 27 (C). The lowest number of isolates, i.e., 18, comes from comparing all plates with one another, regardless of the origin (D). The investigator should make clear which system is used, and although only individual animal/individual organ isolation (A) gives exact data, it should be stated in the examination report which bacteria were found in which animals. In general, grouped animals/individual organs isolation (C) is fully usable for routine examinations.

Colonies are picked from indicative media in the same way, but only colonies with characteristics like those searched for are picked. To facilitate this process it is of great help to grow a plate with a reference strain for each of the organisms to be isolated.

References

1. Møller, V. and Reyn, A., A new solid medium for the isolation of Neisseria gonorrhoeae, *Bull. WHO*, 32, 471, 1965.
2. Gibbons, R.J. and MacDonald, J.B., Hemin and vitamin K compounds as required factors for the cultivation of certain strains of *Bacteroidies melaninogenicus*, *J. Bacteriol.*, 80, 164, 1960.

chapter four

Identification of bacteria

Contents

4.1 Initial characterization of the isolates 55
4.2 Conclusive identification. 56
4.3 Specific techniques used for identification of bacteria 60
      4.3.1   Gram-stainability tests 60
      4.3.2   Other methods used for describing the shape of bacteria 60
      4.3.3   Motility tests 62
      4.3.4   Test for aerobic and anaerobic growth 62
      4.3.5   Catalase test 63
      4.3.6   Cytochrome oxidase test 64
      4.3.7   Acid-fast or spore staining 64
      4.3.8   Carbohydrate fermentation and utilization assays 64
      4.3.9   Disc methods 65
      4.3.10 Other assays 68
      4.3.11 Commercial test kits 68
References 71

4.1 Initial characterization of the isolates (Figure 4.1)

When pure cultures are obtained on nonselective agar plates, each culture should be subjected to some basic tests. First, the Gram stainability and shape of the bacteria have to be characterized as either Gram positive or Gram negative and coccus or rod, respectively. Other basic characteristics are motility, aerobic and anaerobic growth, catalase, cytochrome oxidase, glucose fermentation (acid formation), and whether carbohydrates are utilized fermentatively or oxidatively. For Gram-positive bacteria acid stability and the formation of spores are basic characteristics as well. When these basic characteristics have been verified every isolate should be allocated to one of the groups shown in Tables 4.1 and 4.2, and a full identification may be carried out according to the specific chapters of this book.

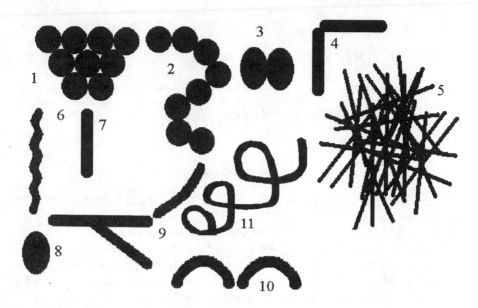

Figure 4.1 Different shapes of bacteria found at microscopy. (1) and (2) are typical Gram-positive cocci (1) grouped as grapes and (2) grouped in chains; (3) are, if Gram positive, also designated as cocci, although more elliptic; as these are grouped as pairs, they are often designated diplococci; (4) is a coryneform; (5) are long, slender rods lying in bundles; (6) is a helical rod; (7) and (8) are typical for Gram-negative rods: (7) is a coliform, while (8) is a pasteurellaform. Gram-negative bacteria must — if designated as cocci — be fully circular as (1) or (2); (9) is a branching, filamentous rod; (10) are curved rods in a pair, appearing as a seagull; (11) forms tight coils or spiral configurations. Laboratory animal bacteria representing the different forms are (1) *Staphylococcus aureus;* (2) *Streptococcus zooepidemicus;* (3) *S. pneumoniae;* (4) *Corynebacterium kutscheri;* (5) *Clostridium piliforme;* (6) *Helicobacter hepaticus;* (7) *Escherichia coli;* (8) *Pasteurella multocida;* (9) *Actinomyces* spp.; (10) *Campylobacter coli;* (11) *Clostridium spiroforme.*

In a particular program for bacteriological screening of laboratory animals, which bacteria to search for should be defined, e.g., as shown in Tables 2.4 to 2.7. After carrying out the basic tests on all isolates, some isolates may have been allocated to groups that do not contain bacteria listed in the program. If the program does not aim at making up a bacteriological profile for the animals, these isolates may be disregarded without any further examination.

4.2   Conclusive identification

This book provides information that should help with the identification of important laboratory animal bacteria. Concerning bacteria being part of the

Table 4.1 First-Stage Table for the Identification of Gram-Positive Bacteria Found in Rodents and Rabbits

	Acid fast	Spores	Motility	Aerobic growth	Anaerobic growth	Catalase	Cytochrome oxidase	Glucose utilization	Carbo-hydrate utilization	Chapter of this book to be used for further identification
Cocci										
Aerococcus	–	–	–	+	+	d	–	+	F	7
Micrococcus	–	–	–	+	–	+	–	d	O/–	7
Peptococcaceae	–	–	–	–	+	–	–	+	F	7
Staphylococcus	–	–	–	+	+	+	–	+	F	7
Streptococcus/ Pediococcus/ Gemella	–	–	d	+	+	–	–	+	F	7
Rods										
Actinomyces	–	–	–	–	+	–	?	d	F	8
Arcanobacterium	–	–	–	+	+	–	–	+	F	8
Bacillus	–	+	d	+	d	+	d	d	F/O/–	8
Clostridium	–	+	d	–	+	–	?	d	F/–	8
Corynebacterium	–	–	–	+	+	+	–	d	–/F	8
Erysipelothrix/ Lactobacillus	–	–	–	+	+	–	–	+	F	8
Kurthia	–	–	+	+	–	+	–	–	–	8

continued

Table 4.1 (continued) First-Stage Table for the Identification of Gram-Positive Bacteria Found in Rodents and Rabbits

	Acid fast	Spores	Motility	Aerobic growth	Anaerobic growth	Catalase	Cytochrome oxidase	Glucose utilization	Carbohydrate utilization	Chapter of this book to be used for further identification
Listeria	-	-	+ (20°C)	+	+	+	-	+	F	8
Mycobacterium	+	-	-	+	?	+	-	+	O/-	9
Nocardia	d	-	-	+	-	+	-	+	O	8
Oerskovia	+	-	+/-	+	+	+	-	+	F	8
Rhodococcus	+	-	-	+	-	+	-	+	O	8
Sporolactobacillus	-	+	-	+	+	-	-	+	F	8

Table 4.2 First-Stage Table for the Identification of Gram-Negative Bacteria Found in Rodents and Rabbits

	Motility	Aerobic growth	Anaerobic growth	Catalase	Cytochrome oxidase	Glucose (acid)	Carbohydrate utilization	Chapter of this book to be used for further identification
Facultative Anaerobes								
Aeromonas	+	+	+	+	+	+	F	10
Enterobacteriaceae	d	+	+	+	-	+	F	10
Haemophilus	-	+	+	+	+[a]	+	F[b]	10
Pasteurella/Actinobacillus	-	+	+	+	+	+	F	10
Streptobacillus moniliformis	-	+	+	-	-	+	F	10

Obligate Aerobes							
Acinetobacter	–	+	+	–	+	O	11
Bordetella	+	+	+	+	–	–	11
Francisella	–	+	ND	ND	ND	ND	11
Flavobacterium	–	+	+	+	+	O	11
Pseudomonas/Agrobacterium/	+	+	+	+	+	O	11
Burkholderia/Sphingomonas							
Weeksella	–	+	+	+	–	–	11
Xanthomonas/Chryseomonas	+	+	+	–	+	O	11
Microaerophilics							
Campylobacter/Helicobacter	+	–c	+	+	–	–	12
Obligate Anaerobes							
Bacteroidaceae	–	–	d	–	d	F/–	12

Note: ND = not properly described as it is not a common procedure in diagnostic laboratories.

a Dependent on the method applied
b Does not grow in Hugh and Leifson's medium
c Grows at microaerophilic incubation

normal flora or the environment of the animals, further information might be found in international indexes for bacteria, the most important of which are *Bergey's Manual of Determinative Bacteriology*,[1] ASM's *Manual of Clinical Microbiology*,[2] and *Cowan and Steel's Bacteriology*.[3] It should be kept in mind that many isolates possess characteristics that will place them between two or more well-defined bacteria, and therefore as such they cannot be fully defined. In particular, within laboratory animal bacteriology, which is still a new and not fully explored field, isolates often belong to groups of bacteria that never have been fully characterized and therefore do not fully fit with definitions of certain bacteria given in *Bergey's Manual*.

4.3   Specific techniques used for identification of bacteria

4.3.1   Gram-stainability tests (Figures 4.2 and 4.3)

Gram stainability can be determined by the traditional Gram staining (Table 4.3) or by the less laborious potassium hydroxide assay: a drop of 3% potassium hydroxide is placed upon the object glass and a rich amount of bacterial mass is mixed into it. Due to hydrolysis of the cell wall, the DNA of a Gram-negative bacterium will form a mucoid immersion and long threads (3 to 5 mm) can be drawn from the solution with the platin needle after 15 to 20 s (Figure 4.4). A Gram-positive bacterium will form an aqueous immersion, from which no threads can be drawn. Commercial strips (Bactident®, Merck, Germany) may be used for testing for-aminopeptidase activity, an enzyme normally found only in Gram-negative bacteria. A strip dipped into an Eppendorf tube with a mixture of water and bacteria is incubated in a 37°C water bath for 10 min. If the mixture turns yellow the bacterium is Gram negative. If no color is evolved the mixture is reincubated for 20 min. If there is still no change in color the bacterium is Gram positive. If there is only a weak change, the assay result is equivocal. As neither test is 100% reliable it is often necessary to combine two or all three tests.

4.3.2   Other methods used for describing the shape of bacteria

Aqueous preparations may be directly microscoped. A drop of water is placed upon the object glass and two to three colonies are mixed into the water, and a cover glass is placed upon the mixture. Immersion oil is placed upon the cover glass and the preparation is microscoped at 1000 × magnification (oil linse). The bacteria are characterized according to Figure 4.1.

Alternatively, a colony may be mixed into India ink on an object glass. The mixture is spread in a thin layer and left until fully dry. The slide is microscoped as a Gram-stained slide. The microscoped field is dark, while the bacteria are seen as "holes" in the dark field.

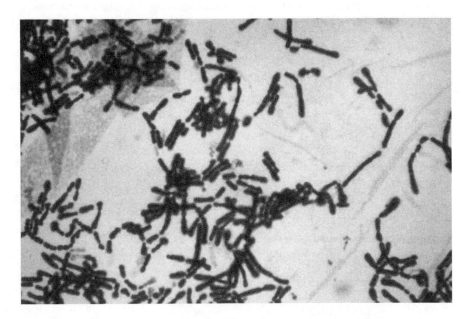

Figure 4.2 Gram-positive bacteria (*Bacillus* spp.)

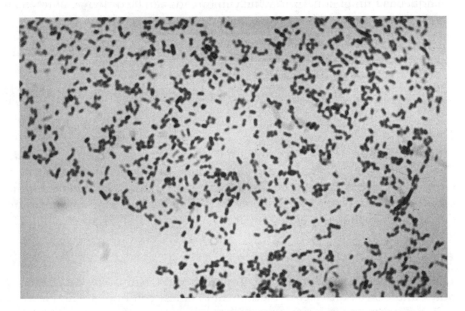

Figure 4.3 Gram-negative bacteria (*Pasteurella pneumotropica*)

Table 4.3 A Method for Gram Staining Bacteria

Materials

Object glass	Weak carbol fuchsin
Platin needle	Microscope
Crystal violet	Flame
Lugol's solution	Immersion oil
96% ethanol	

Preparation

A drop of water is placed on the object glass and two to three colonies are mixed into the water and spread over the surface of the glass. The glass is air dried and flame fixed.

Method

1. Crystal violet is spread over the surface of the glass and left for 1 min.
2. The glass is washed with Lugol's solution.
3. Lugol's solution is spread over the surface of the glass and left for 1 min.
4. The glass is washed with ethanol until the blue color disappears, at least outside the area where the bacteria have been spread.
5. The glass is washed with water.
6. Carbol fuchsin is spread over the surface and left for 15 to 20 s.
7. The glass is washed with water. A piece of filter paper is pressed onto the surface, and the glass is left until fully dried.

Microscopy of the Gram-stained slides

A drop of oil is placed directly upon the surface of the glass without using a cover glass. Microscopy is done with an oil linse at 1000 × magnification. The shape of the bacteria is characterized according to Figure 4.1. Blue bacteria are Gram positive (Figure 4.2); red bacteria are Gram negative (Figure 4.3).

4.3.3   Motility tests

Motility can be diagnosed by direct microscopy of a droplet from a broth. The bacteria should be moving in all directions — not just in one direction, which may be the cause of nonmotile bacteria by the waterstreams on the slide. Alternatively, the bacteria may be inoculated in a semisolid medium (Table 4.4), which is inoculated and read as described in Figure 4.5.

4.3.4   Test for aerobic and anaerobic growth

Testing for aerobic and anaerobic growth is done simply by incubating a pure culture on a nonselective medium under these conditions. Alternatively, a high agar, i.e., an appropriate medium in a high tube, may be inoculated and interpreted as described in Figure 4.6.

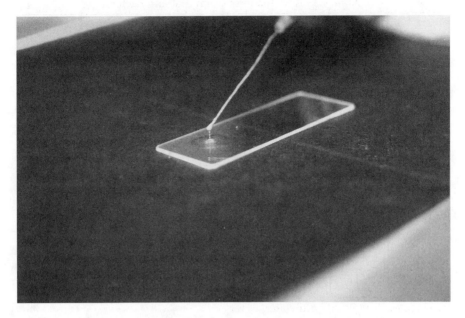

Figure 4.4 Drawing long threads from a Gram-negative bacterium after immersion into potassium hydroxide.

Table 4.4 A Semisolid Medium
for Testing Motility of Bacteria

Gelatine	80 g
Distilled water	1000 ml
Peptone	10 g
Beef extract	3 g
Sodium chloride	5 g
Agar	4 g

4.3.5 Catalase test

Test for catalase activity is performed by placing some colony mass on a slide and dripping 3% hydrogen peroxide onto the mass. If small bubbles develop — which may only be observed under a microscope — the test is positive.

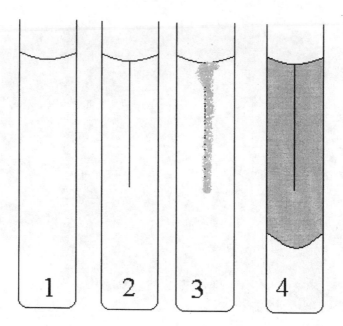

Figure 4.5   Inoculation of a semisolid medium for determination of motility of bacteria. The medium is only inoculated halfway from top to bottom (2). The medium is then incubated at 37°C, and should be read after 6 to 8 h incubation, and again after 18 to 24 h. If the bacterium is nonmotile, it will only grow along the inoculation channel (3), while motile bacteria will spread from the channel in all directions (4).

4.3.6   Cytochrome oxidase test

Testing for cytochrome oxidase activity is performed by dripping oxydase reagent (Table 4.5) onto some colony mass on a filter paper. A positive reaction is shown as a deep blue color within 10 s. It should be noted that this method, originally described by Kovács,[4] is more sensitive than a method later described by Gaby and Hadley.[5] Therefore, in some international indexes, *Haemophilus* spp. are described as oxidase positive, while in other indexes they are described as oxidase negative.

4.3.7   Acid-fast or spore staining

Acid-fast bacteria are resistant to destaining with sulfuric acid. A method for staining is shown in Table 4.6. This method may be modified to stain spores.

4.3.8   Carbohydrate fermentation and utilization assays

Basically, carbohydrate fermentation is tested in a broth supplemented with a specific carbohydrate and an indicator for acid production. A recipe is given in Table 4.7. If test for gas production is necessary a so-called Durham

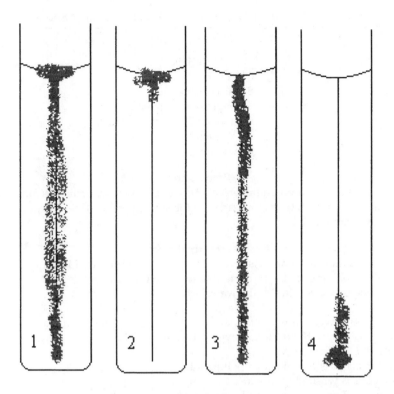

Figure 4.6 High agar inoculation. A bacterial isolate has been inoculated from top to bottom in a tube and then incubated for 24 h. 1 is facultatively anaerobic, 2 is obligate aerobic, 3 is microaerophilic, while 4 is obligate anaerobic.

Table 4.5 Kovács' Reagent for Testing Cytochrome Oxidase Activity

Tetramethyl-*p*-phenylene-diamine	1 g
Ascorbic acid	1 g
Distilled water	1000 ml

tube, an inverted glass tube, is placed in the bottom of the broth. Whether the utilization of the carbohydrates is fermentative or oxidative should be tested in a medium described by Hugh and Leifson.[6]

4.3.9   Disc methods

Testing for sensitivity to antibiotics is done by agar diffusion inhibition assay, in which a disc with a minor amount of the antibiotic is placed upon an agar plate with the isolate. After a cultivation period an inhibition zone around

Table 4.6 A Method for Staining Acid-Fast Bacteria or Spores

Materials
Object slides
Burning flame
Strong carbol fuchsin
Ethanol or 5% sulfuric acid
1% malachite green
Platin needle

Preparation
A drop of water is placed on the object glass and two to three colonies are mixed
into the water, which then is spread over the surface of the glass. The glass is air
dried and flame fixed.

Method
1. The carbol fuchsin solution is spread over the surface of the slide, which is then
   heated to boiling over the burning flame and left for 1 to 5 min.
2. The slide is washed under the water tap.
3. *Spores*
   The glass is washed with ethanol until the color disappears, at least outside
   the area where the bacteria have been spread.
   Acid-fast bacteria
   Sulfuric acid is used instead of ethanol.
4. The glass is washed with water.
5. Steps 3 to 5 are repeated until the smear is fully decolorized.
6. Malachite green is spread over the surface and left for 30 s.
7. The glass is washed with water. A piece of filter paper is pressed onto the
   surface, and the glass is left until fully dried.
8. The slide is microscoped as described for Gram-stained slides (Table 4.3).
   Spores
   Vegetative bacteria are green, while the spores are red. These may either be
   situated alone or inside a bacterial cell. If so, they should be characterized as
   described in Figure 8.1.
   Acid-fast bacteria
   Acid-fast bacteria are red, while other bacteria are green.

the disc is measured. The concentration of the antibiotic decreases with
increasing distance from the disc, dependent on the diffusion constant of the
antibiotic.Therefore, the cutoff radius differs between different antibiotics.
In practice, two cutoff radii are used, one below which the isolate is said to
be *intermediate* resistant to the antibiotic, and a shorter radius, below which
the isolate is defined as *resistant*. Radii higher than both cutoff radii indicate
that the isolate is *sensitive*.

Table 4.7 A Broth for Testing Carbohydrate
Fermentation

Beef extract	5.0 g
Bromthymol blue	0.024 g
Carbohydrate	5.0 g
Disodium-hydrogene phosphate	2.0 g
Peptone	10.0 g
Sodium chloride	3.0 g

Note: After inoculation the broth should be incubated at
37°C for up to 1 week and inspected daily. A change
in color from blue to yellow is interpreted as positive.

Table 4.8 A Recipe for Mueller Hinton Agar To Be Used
for Agar Diffusion Inhibition Assays[a]

Water	1000 ml
Peptone 5 (Acid digest of caseine)	21.5 g
Beef infusion	2.0 g
Agar	13.0 g

Note: The mixture is boiled to dissolve, and then autoclaved. After
cooling to 50°C, 5% v/v horse or sheep blood may be added
if necessary for growth. The pH should be adjusted to 7.1 to
7.5.

[a] The agar base is commercially available from Gibco (U.S.).

Data from Bauer, A. W., Kirby, W. M., Sherris, J. C., and Turck, M.,
Am. J. Clin. Pathol., 45(4), 493, 1966.

The assay is performed on Mueller Hinton agar (Table 4.8). The antibiotic
discs may either be prepared in the laboratory as paper discs soaped in an
antibiotic solution or bought from commercial suppliers, e.g., Rosco (Denmark). They should be placed upon the agar with an automatic dispenser.

The isolate is suspended in 1 ml sterile distilled water to make a density
of McFarland 0.5. The suspension is whirl-mixed. A sterile cotton stick is
dipped into the suspension several times, then surplus suspension is pressed
out of the stick by pressing it against the inside of the tube. The cotton stick
is then used for streaking upon the agar surface, which is done three times
leaving no part of the surface untouched. Finally, the entire edge of the agar
is touched by the stick all the way around. The agar is left for drying 3 to
15 min, and the disks are placed with the dispenser. After 16 to 18 h of
incubation at 37° (staphylococci and enterococci should be incubated 24 h),

Table 4.9 Presumptive Inhibition Zones for Testing Antibiotic Sensitivity
by the Use of Antibiotic Discs on Mueller Hinton Agar[a]

	Diameter (mm) of inhibition zone	
	Sensitive	Intermediate
Ampicillin	19	16
Amoxicillin with clavulanic acid	19	16
Penicillin	25	25
Cephalosporins	27	23
Tetracycline	22	22
Chloramphenicol	25	22
Fucidine	27	23
Erythromycin	25	19
Tylosin	25	22
Spiramycin	25	22
Neomycin	22	19
Spectinomycin	19	16
Streptomycin	25	22
Lincomycin	25	22
Tiamuline	12	10
Sulfonamide	22	19
Trimethoprim	19	16
Sulfonamide with trimethoprim	25	22
Enrofloxacin	19	16
Gentamicin	22	19

[a] Neo Sensitabs®, Rosco (Denmark).

the inhibition zones are read and interpreted according to the cutoff values
given by the producer of the discs. As a rule of thumb the cutoff values in
Table 4.9 may be used.

4.3.10 Other assays

Several biochemical assays are needed in a laboratory for full identification
of all important laboratory animal bacteria. Describing every assay in detail
goes beyond the scope of this book. Descriptions may be found in textbooks
on general bacteriology, e.g., References 2 and 3, or instructions for specific
tests may be found in the references given in Table 4.10.

4.3.11 Commercial test kits

To reach a final conclusion on the identity of an isolated bacterium, a certain
range of tests is necessary. Commercial kits containing a number of tests,
which should lead to the full identification of bacteria belonging to the group
for which the kit has been designed, represent an alternative to using a range

Table 4.10 Important Assays for Identification of Bacteria

Characteristic shown by the assay	Ref.
Aminoacid decarboxylase activity	8
Bile sensitivity	9
CAMP test	3, 10
Citrate utilization	11
Coagulase activity	12
Deoxyribonuclease activity	13, 14
Gelatine liquefaction	3, 15, 16
Hydrogen sulfide production	17, 18
Indole reaction (tryptophanase activity)	19
Lecithinase and lipase activity	3
Malonate utilization	20
Nitrate reduction	21
ONPG test (β-galactosidase activity)	22, 23
PGUA test (β-glucuronidase activity)	24
PNPX test (β-xylosidase activity)	24
Phenylalanine deaminase activity	25
Porphyrin formation	26
Potassium cyanide resistance	27
Tartrate utilization	28
Tellurite resistance	29
Urease activity	31
Voges-Proskauer test (acetoin formation)	32

Note: Instructions may be found in either the references or in *Cowan and Steel's Manual for the Identification of Medical Bacteria.*[3]

of in-house tests. Some of these kits are for manual inoculation, reading, and interpretation, while others are partly or fully automatized.

One particular system, the API system (Figure 4.7) sold by bioMérieux, France, has been widely used by laboratory animal bacteriologists due to its wide range of applications. It consists of a range of strips, each of which contains from 10 to 50 different substrates. Each strip is designed for a certain group of bacteria. The strips are inoculated according to a given instruction and then reactions are read; some tests demand the addition of a reagent prior to reading. Some kits are readable by an automatic reader. The results all together lead to a numeric profile, which is computerized to give the identity of the organism tested along with a probability of a correct diagnosis. The manual and the computer software base their identification on a percentage of standard results given by the manufacturer for each test and each bacterium, e.g., in test for urease 85% of *Pasteurella pneumotropica* isolates react positively.

Commercial systems are extremely valuable for the laboratory animal bacteriologist. Table 4.11 lists the names of API strips suitable for important

Figure 4.7   API 20E commercial strips (bioMérieux, France) for the identification of Gram-negative bacteria.

Table 4.11   API kits (bioMérieux, France) Suitable
for the Diagnosis of Important Laboratory
Animal Bacteria

	API kit
Bordetella bronchiseptica	20 NE
Citrobacter rodentium	20 E
Campylobacter spp.	Campy
Corynebacterium kutscheri	Coryne
Erysipelothrix rhusiopathiae	Coryne
Haemophilus spp.	NH
Helicobacter spp.	Campy
Pasteurella spp.	20 NE
Salmonellae	20 E
β-Hemolytic streptococci	20 Strep
Streptococcus pneumoniae	20 Strep
Yersinia pseudotuberculosis	20 E

bacteria in laboratory animals. However, limitations on the use of such systems within laboratory animal bacteriology should be noted and one should never rely solely on a commercial kit identification. The following points are important:

1. The kits are designed for human use. Profiles of the bacteria included in the manuals or computer programs are not always directly comparable with animal strains of the same bacteria. For example, if rat strains of a certain bacterium differ from human strains in a certain test, this usually will not be listed or will be listed only as a low percentage reaction. Some bacteria, e.g., the murine *Citrobacter rodentium*, are not included in the API system, and therefore the profile must be known from other sources.
2. The individual test results often may be used to make up a diagnosis without the use of the commercial manual or computer software. However, occasionally alternative indicators have been applied, which may lead to alternative results. If this is the case, a standard in-house test should be applied in the laboratory as a supplement.

To avoid false diagnoses as a result of using commercial kits, a diagnosis on a generic level should always be reached by the use of basic standard tests before attempting full identification with a commercial kit. Furthermore, quality assurance of the kits should be performed by screening reference strains of laboratory animal bacteria in the kits used in the laboratory, and the profiles obtained by this procedure should be used along with manuals and computer software.

References

1. Holt, J. G., Krieg, N. R., Sneath, P. H. A., Staley, J. T., and Williams, S. T., *Bergey's Manual® of Determinative Bacteriology*, 9th ed., William & Wilkins, Baltimore, 1994.
2. Murray, P. R., Baron, E. J., Pfaller, M. A., Tenover, F. C., and Yolken, R. H., *Manual of Clinical Microbiology*, ASM Press, Washington, D.C., 1995.
3. Barrow, G. I. and Feltham, R. K. A., *Cowan and Steel's Manual for the Identification of Medical Bacteria*, Cambridge University Press, Cambridge, 1993.
4. Kovács, N., Identification of *Pseudomonas pyocyanea* by the oxidase reaction, *Nature*, 178, 103, 1956.
5. Gaby, W. L. and Hadley, C., Practical laboratory tests for *Pseudomonas aeruginosa*, *J. Bacteriol.*, 74, 356, 1957.
6. Hugh, R. and Leifson, E., The taxonomic significance of fermentative versus oxidative metabolism of carbohydrates by various Gram-negative bacteria, *J. Bacteriol.*, 66, 24, 1953.
7. Bauer, A. W., Kirby, W. M., Sherris, J. C., and Turck, M., Antibiotic susceptibility testing by a standardized single disk method, *Am. J. Clin. Pathol.*, 45(4), 493, 1966.
8. Møller, V., Simplified tests for some amino acid decarboxylases and for the arginine dehydrolase system, *APMIS*, 36, 158, 1955.
9. MacConkey, A. T., Bile salt media and their advantages in some bacteriologic examinations, *J. Hyg. (Cambridge)*, 8, 322, 1908.

10. Koser, S. A., Utilization of salts of organic acids by the colon-aerogenes group, *J. Bacteriol.*, 8, 423, 1923.
11. Christie, R., Atkins, N. E., and Munch-Petersen, E., A note on a lytic phenomenon shown by group B Streptococci, *Austr. J. Exp. Biol. Med. Sci.*, 22, 197, 1944.
12. Cowan, S. T., The classification of Staphylococci by precipitation and biological reactions, *J. Path. Bact.*, 46, 31, 1938.
13. Jeffries, C. D., Holtman, D.F., and Guse, D.G., Rapid method for determining the activity of microorganisms on nucleic acids, *J. Bacteriol.*, 73, 5901, 1957.
14. Lachica. R. V. F., Genigeorgis, C., and Hoeprich, P. D., Metachromatic agar-diffusion methods for detecting staphylococcal nuclease activity, *Appl. Microbiol.*, 21, 585, 1971.
15. Kohn, J., A preliminary report of a new gelatin liquefaction method, *J. Clin. Pathol.*, 6, 249, 1953.
16. Sierra, G., A simple method for the detection of lipolytic activity of microorganisms and some observations on the influence of the contact between cells and fatty substrates, *Antonie v Leeuwenhoek*, 23, 15, 1957.
17. Kolmos, H. J. and Schmidt, J., Failure to detect hydrogen-sulphide production in lactose/sucrose-fermenting *Enterobacteriaceae*, using triple sugar iron agar, *APMIS* Sect. B, 95, 85, 1957.
18. Clarke, P. H., Hydrogen sulphide production by bacteria, *J. Gen. Microbiol.*, 8, 397, 1953.
19. Böhme, A., Die Anwendung der Ehrlichsen Indolreaktion für bakteriologische Zwecke [The application of Ehrlich's indole reaction in bacteriology], *Zbl. Bakt. I. Abt. Orig.*, 40, 129, 1905.
20. Leifson, E., The fermentation of sodium malonate as a means of differentiating *Aerobacter* and *Escherichia*, *J. Bacteriol.*, 26, 329, 1933.
21. Zobell, C. E., Factors influencing the reduction of nitrates and nitrites by bacteria in semisolid media, *J. Bacteriol.*, 24, 273, 1932.
22. Bülow, P., The ONPG test in diagnostic bacteriology. I. Methodological investigations, *APMIS*, 60, 376, 1964.
23. Bülow, P., The ONPG test in diagnostic bacteriology. I. Comparison of the ONPG test and the conventional lactose fermentation test, *APMIS*, 60, 387, 1964.
24. Kilian, M. and Bülow, P., Rapid diagnosis of *Enterobacteriaceae*. I. Detection of bacterial glycosidases, *APMIS* Sect. B, 84, 245, 1976.
25. Berquist, L. M. and Searcy, R. L., A micromethod for the detection of utilization of phenylalanine by microorganisms, *Amer. J. Clin. Pathol.*, 39, 544, 1963.
26. Kilian, M., A rapid method for the differentiating *Haemophilus* strains. The porphyrin test, *APMIS* Sect. B, 82, 835, 1974.
27. Møller, V., Diagnostic use of the Braun KCN test within the *Enterobacteriaceae*, *APMIS*, 34, 115, 1954.
28. Kauffmann, F. and Petersen, A., The biochemical group and type differentiation of *Enterobacteriaceae* by organic acids, *APMIS*, 38, 481, 1956.
29. Anderson, J. S., Happold, F. C., McLeod, J. W., and Thomson, J. G., On the existence of two forms of diphteria bacillus, *B. diphteria gravis* and *B. diphteria mitis* — and a new medium for their differentiation and for the bacteriological diagnosis of diphteria, *J. Pathol. Bacteriol.*, 34, 667, 1931.
30. Christensen, W. B., Urea decomposition as a means of differentiating *Proteus* and paracolon cultures from each other and from *Salmonella* and *Shigella* types, *J. Bacteriol.*, 52, 461, 1946.

31. Stuart, C. A., van Stratum, E., and Rustigan, R., Further studies on urease production by Proteus and related organisms, *J. Bacteriol.*, 49, 437, 1945.
32. O'Meara, R.A.Q., A simple delicate and red method of detecting the formation of acetylmethyl carbinol by bacteria fermenting carbohydrate, *J. Pathol. Bacteriol.*, 34, 401, 1931.

chapter five

Immunological methods

Contents

5.1   Agglutination..76
5.2   Immunofluorescence techniques...77
    5.2.1   Diagnosing the presence of bacteria in a sample 77
    5.2.2   The immunofluorescence assay 81
5.3   Immunoenzymatic staining 82
5.4   Enzyme-linked immunosorbent assay 86
    5.4.1   Principles...86
    5.4.2   The microtiter plates 87
    5.4.3   The antigen...93
    5.4.4   Antibodies, enzymes and substrates..................................93
    5.4.5   Coating the wells..94
    5.4.6   Blocking the wells 94
    5.4.7   Performing the assay 95
    5.4.8   Control sera 95
    5.4.9   Interpretation of the OD-value..97
5.5   Immunoblotting 98
References..99

Immunological methods, i.e., methods based on an antigen-antibody reaction, are used in two ways: *antigen detection*, in which specific antibodies directed against a certain agent are used to show the presence of this agent in a sample or identify a pure culture as the agent; and *serology*, in which the animal by the use of the agent or parts of it as antigen is shown to possess antibodies against that agent.

Immunological identification may be applied to a sample in which the presence of a certain bacterium is suspected. This requires either a high amount of the agent in the sample or a high sensitivity of the method, although a lower sensitivity may be accepted if identification is applied to pure cultures. Setting up the method, e.g., when pretesting the antibodies, attention is normally paid to avoiding cross-reactions with closely related bacteria, while bacteria with a more distant relationship would not be tested. One should, therefore, never try to identify a pure culture by an immunological method if the culture has not been subjected to basic tests, and these tests have allocated the isolate to the group including that bacterium against which the antibody is directed.

If antigen detection is used to diagnose an infection it will often be sufficient to immunostain a smear prepared from the infected organ. Pathologists might have an interest in describing the position of the antigen inside the tissue, and, therefore, they often stain slides prepared as for histopathology, a technique that is not covered by this book.

Serology is the art of identifying the presence of antibodies in serum. For those bacteria for which cultivation methods are too insensitive or impossible, e.g., *Clostridium piliforme* or *Helicobacter hepaticus*, the demonstration of antibodies may be a more reasonable approach. The main pitfalls are the lack of sensitivity due to a pure antibody response to the infection or a low specificity due to cross-reactions, as typical bacteria contain more than 2000 antibody-inducing proteins. *C. piliforme*, CAR bacillus, *Mycoplasma* spp., and *Leptospira* spp. are quite easily diagnosed by serology, as they induce a high amount of serum IgG.

5.1   Agglutination

The simplest immunological method is to agglutinate the antigen by a serum containing specific antibodies. The presence of antibodies in the serum is shown by positive agglutination. Alternatively, if the antigen is the unknown factor, the positive agglutination indicates the identity of the antigen. The method may be performed either in a tube, on a slide, or on a piece of black cardboard. The visualization of the agglutination on a slide may be enhanced if the known factor, i.e., either the bacterial antigen or the specific antibodies are linked to polystyrene (latex) particles, which are generally the size of 1 μ. The method is not very sensitive as a tool for diagnosing antibodies, and in laboratory animal bacteriology it is mainly used for diagnosing antibodies to *Leptospira* spp., as antigen-coated latex particles are commercially available from Sanofi Pasteur (U.S.). For diagnosing the presence of certain bacteria in a nonpurified sample this method is normally too insensitive to be of any practical use, but as these tests require a minimum of time and skill for the performance, they are extremely valuable for a quick identification of a pure culture.

The technique is commonly applied for the quick identification of bacteria. Many bacteria will react directly with the antibodies, while the reaction of others may be enhanced if the bacteria have been digested by an enzyme. Latex-linked antibodies against various bacteria, e.g., *Staphylococcus aureus* and Lancefield's groups of streptococci, are commercially available from bioMérieux (France) or Meridian Diagnostics (U.S.).

5.2 Immunofluorescence techniques

5.2.1 Diagnosing the presence of bacteria in a sample

The principles of immunofluorescence techniques are shown in Figure 5.1. A specific antibody, preferably a monoclonal antibody, is used for staining of bacteria. The antibody should be pretested in different dilutions. The antibody is subsequently used in a dilution that is considerably lower than the highest dilution given fluorescence at pretesting. An example of a protocol for indirect immunofluorescence staining is shown in Table 5.1 (Also see Figures 5.2 through 5.4 pertaining to this table.) Samples to be tested can simply be smeared on a Teflon-coated slide (Figure 5.2) and fixed by the use of –20°C acetone. Formalin should not be used as this may cover the antigens for the antibody. Immunofluorescence staining is, in general, more sensitive than peroxidase-antiperoxidase staining (see below). In laboratory animal bacteriology the method has been applied as a simple tool for staining of *Clostridium piliforme* (Figure 5.5). Fluorescein-conjugated, species-specific anti-immunoglobulins are available from several commercial suppliers. Dako Corporation (U.S.), Charles River (U.S.), Harlan (U.K.), and Organon Teknika (U.S.). The

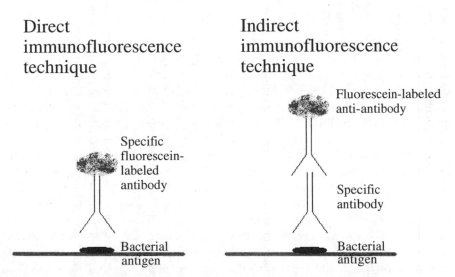

Direct
immunofluorescence
technique

Specific
fluorescein-
labeled
antibody

Bacterial
antigen

Indirect
immunofluorescence
technique

Fluorescein-labeled
anti-antibody

Specific
antibody

Bacterial
antigen

Figure 5.1   The principle in immunofluorescence techniques. The bacterial antigen is linked to a glass slide. If a specific fluorescein-labeled antibody is available the bacteria are directly visualized. If such an antibody is not available, or the possible detection of antibodies in a serum sample is the aim, the indirect technique is used: the specific antibody reacts with the bacterial antigen, and a fluorescein-labeled antiantibody reacts with the specific antibody. In both techniques, subsequent microscoping in a fluorescence microscope will show fluorescent bacteria (Figure 5.5).

Table 5.1   Example of a Protocol for Immunofluorecence Staining for the Identification of Bacteria in Organ Samples

Materials

8- or 10-well Teflon slides
Arced tweezers
Acetone (–20°C)
Scissors
Filter paper
Phosphate-buffered saline (PBS)
Micropipette
Primary antiserum: Antiserum against the agent to be stained

Conjugate antiserum: Fluorescein-labeled antiserum against the immunoglobulin of the animal species delivering the primary antiserum
Humid chambers (Figure 5.2)
Washing trunks (Figure 5.3)
Cotton sticks
Filter paper
Fluorescense microscope (Figure 5.4)
Mounting fluid (e.g., 9:1 PBS glycerol)

Preparation

Tissue samples: A piece of the organ is sampled. With the arced tweezers the cut surface is pressed gently against a piece of filter paper to remove excess blood and a thin smear is made over the wells of the Teflon slides (cut surface toward the slide). The Teflon slides are left at room temperature until dry, then covered with –20°C acetone (use a safety cabinet!), and again left until dry. If not stained immediately the slides may be frozen at –20°C.

Pure cultures: A broth or a suspension is diluted until one drop contains several but not intermingling bacteria. One drop (10 to 15 μl) is placed in each Teflon slide well. The Teflon slides are left at room temperature until dry, then covered with –20°C acetone (use a safety cabinet!), and again left until dry. If not stained immediately the slides may be frozen at –20°C.

Staining

1. Dilute the primary antiserum and the conjugate antiserum in a proper dilution found at pretesting.
2. Place the slides in the humid chamber filled with water.
3. Put 12 μl diluted primary antiserum onto each well. Close the chamber and incubate at room temperature for 30 min.
4. Wash three times 5 min each time in PBS in the washing trunks.
5. Dry up all fluid between the wells by the use of the cotton sticks. Do not dry the wells.
6. Put 12 μl diluted conjugate antiserum onto each well. Close the chamber and incubate it at room temperature for 30 min.
7. Wash three times 5 min each time in PBS in the washing trunks.
8. Leave the slides until dry.
9. Place a tiny drop of mounting fluid on each well. Avoid bubbles. Place a cover glass on the slide.
10. Perform fluorescence microscopy at 400 × or 1000 × (immersion oil) magnification.

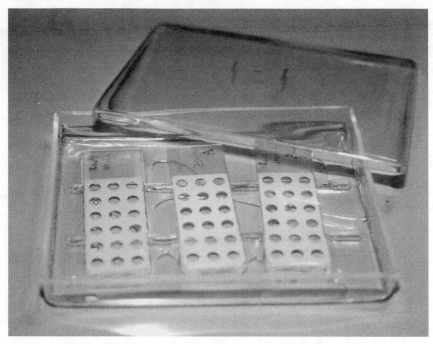

Figure 5.2　Teflon-coated slides for immunofluorescence staining placed in a humid chamber.

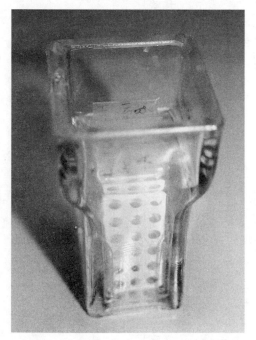

Figure 5.3　Washing trunks for washing immunofluorescence slides.

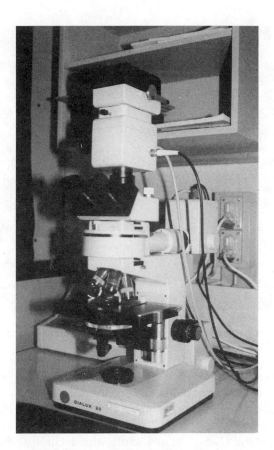

Figure 5.4   Fluorescence microscope.

method may as well be applied on pure cultures for identification of these with specific antisera, as it may be done with mycoplasmas.

5.2.2   The immunofluorescence assay

Using immunofluorescence techniques for demonstrating antibodies (serology) is in principle the same as demonstrating the antigen (see Figure 5.1). For demonstrating antibodies only the indirect technique is usable, referred to as the *immunofluorecence assay (IFA)*. This method is mostly applicable for diagnosing antibodies to bacteria of a certain size, as during the microscopic evaluation it may be difficult to differentiate small bacteria from unspecific fluorescence. If the bacterium can be grown in superficial media, it is rather simple to inoculate a broth, incubate it, and dilute it until the bacterial density is acceptable for microscopy. A drop of the diluted broth is then put onto a Teflon-coated slide and fixed with –20°C acetone. For bacteria such as *C. piliforme*, which have to be harvested from experimental infections, a

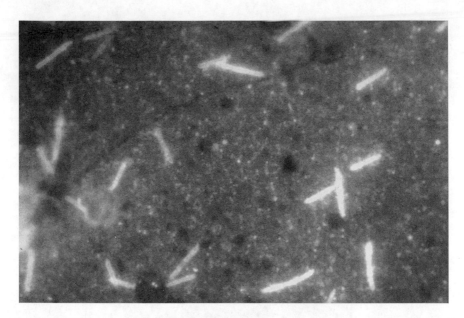

Figure 5.5 *Clostridium piliforme* stained by indirect immunofluorescence technique (Courtesy of Frederik Dagnœs-Hansen).

thin smear of the infected organ has to be made as described in Table 5.1. For some agents antigen-coated slides are commercially available; for example, *C. piliforme*-coated slides may be purchased from Harlan Sprague-Dawley (U.S.), while *Treponema cuniculi*-coated slides may be purchased from Sanofi Pasteur (U.S.). In principle, the technique is performed in exactly the same way as described for staining slides in Table 5.1. The test serum is diluted prior to testing. Different laboratories use different dilutions, e.g., 1:10,[1] 1:20,[2] or 1:40.[3] It is recommended to raise the specificity of the test by using a test dilution of 1:40 or even more in order to avoid false positives and to compensate for the reduced sensitivity by testing more samples, as described in Chapter 1.

5.3 Immunoenzymatic staining

As an alternative to using fluorescein-conjugated antibodies for staining bacteria, antibodies conjugated with enzymes, such as peroxidase or alkaline phosphatase, may be used. The principle is shown in Figure 5.6. The slides can be microscoped in an ordinary light microscope, which facilitates a better visualization of the *in situ* placement of the bacterium and makes immunological staining available to those lacking a fluorescence microscope. A clear disadvantage of the enzymatic technique is a lower sensitivity, which,

One-step direct method

Two-step indirect method

Enzyme-antienzyme method

Labeled avidin-biotin method

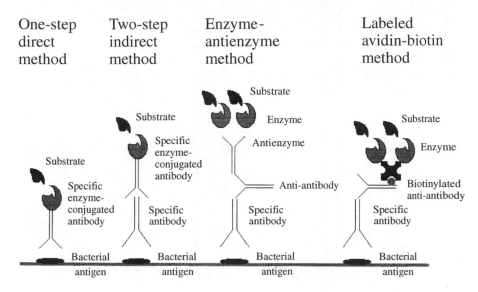

Figure 5.6 Immunoenzymatic staining methods. The antigen may be stained directly by the use of an enzyme-conjugated antibody specifically directed against the antigen. To raise the sensitivity various indirect methods may be applied by using antienzymes or the avidin-biotin system.

however, may be counteracted by using the enzyme-antienzyme technique (Figure 5.6). This method is more sensitive due to the presence of more enzyme molecules. Staining with peroxidase, referred to as PAP staining, is generally usable, but in some specimens a high amount of endogenous peroxidase is present and under such circumstances alkaline phosphatase should be used instead — so-called APAAP staining. A protocol for PAP staining is shown in Table 5.2. Another way of raising the sensitivity is to use the proteins avidin or streptavidin along with the coenzyme biotin. Avidin or streptavidin have a high affinity for biotin, i.e., if a biotinylated antibody is used as the secondary anti-antibody, this may be visualized by the use of enzyme-labeled avidin molecules.

The substrates used for ELISA (see below) should not be used for this technique, as substrates should not be water soluble as they need to be for ELISA. For horseradish peroxidase either 3'-diaminobenzidine tetrahydrochloride (DAB) or 3-amino-9-ethylcarbazole (AEC) is used, while fast red or new fuchsin is used for alkaline phosphatase. All are commercially available from DAKO Corporation (U.S.).

Immunoenzymatic staining may be applied for staining a wide range of bacteria. Quick-staining kits only to be supplied with a specific antibody are commercially available from DAKO Corporation (U.S.).[4]

Table 5.2 A Suggested Protocol for Peroxidase–Antiperoxidase–Staining

Materials

Washing bottle with distilled water

PBS

Washing trunk

Cotton sticks

Arced tweezers

Rabbit serum (dilution 1:20 in PBS)

Primary antibody (specific antibody to the agent in question, diluted 1:40 in PBS)

Humid chamber

Mounting fluid (e.g., 9:1 PBS glycerol)

Microscope

Anti-immunoglobulin (specific for the species delivering the primary antibody, diluted 1:40 in PBS)

Horseradish peroxidase-antiperoxidase (PAP) complex (antibody must be specific for the species delivering the primary antibody, diluted 1:40 in PBS)

DAB tablets

0.05 M Tris buffer, pH 7.6

3% hydrogen peroxide

Preparation

A piece of the organ is sampled. With the arced tweezers the cut surface is pressed gently against a piece of filter paper to remove excess blood and a thin smear is made over the wells of the Teflon slides (cut surface toward the slide). The Teflon slides are left at room temperature until dry, then covered with –20°C acetone (use a safety cabinet!), and again left until dry. If not stained immediately the slides may be frozen at –20°C.

Preparation of the DAB substrate solution

Dissolve 6 mg DAB in 10 ml of the Tris buffer and add 0.1 ml hydrogen peroxide (should be filtered if precipitate forms). Solution is stable for 2 h at room temperature.

Method

1. Gently rinse the slide with distilled water from a wash bottle.
2. Place slide in PBS in the washing trunk.
3. Use a cotton stick for removing excess liquid from around the specimen.
4. Apply 4 to 6 drops of normal rabbit serum (dilution 1:20).
5. Tap off serum and wipe away excess.
6. Apply 4 to 6 drops of the primary antibody (dilution 1:40).
7. Incubate in a humid chamber for 30 min.
8. Repeat steps 1 to 5.
9. Apply 4 to 6 drops of anti-immunoglobulin (dilution 1:40).
10. Incubate in a humid chamber for 30 min.
11. Repeat steps 1 to 5.
12. Apply 4 to 6 drops of horseradish antiperoxidase-rabbit(-mouse)-antiperoxidase (PAP) complex (dilution 1:40).
13. Incubate in a humid chamber for 30 min.
14. Gently rinse the slide with distilled water from a wash bottle.
15. Use a cotton stick for removing excess liquid from around the specimen.
16. Apply DAB substrate solution to give a colored end product and incubate until desired color intensity has developed.
17. Gently rinse the slide with distilled water from a wash bottle.
18. Leave the slides until dry.
19. Place a tiny drop of mounting fluid on each well. Avoid bubbles. Place a cover glass on the slide.
20. Perform microscopy at 400 × or 1000 × (immersion oil) magnification.

5.4   Enzyme-linked immunosorbent assay

5.4.1   Principles

The principles of *enzyme-linked immunosorbent assay (ELISA)* are shown in
Figure 5.7. ELISA is usually performed in a 96-well microtiter plate (Figure

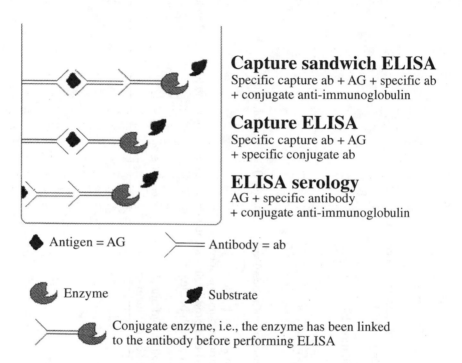

Capture sandwich ELISA
Specific capture ab + AG + specific ab
+ conjugate anti-immunoglobulin

Capture ELISA
Specific capture ab + AG
+ specific conjugate ab

ELISA serology
AG + specific antibody
+ conjugate anti-immunoglobulin

◆ Antigen = AG   Antibody = ab

Enzyme   Substrate

Conjugate enzyme, i.e., the enzyme has been linked
to the antibody before performing ELISA

Figure 5.7   Enzyme-linked immunosorbent assay (ELISA). The assay is performed
in a multiwell microtiter plate (Figure 5.8). Antigen-antibody reaction is visualized
by an enzymatic reaction, in which a colored product develops from a substrate. This
is read photometrically and quantified as the reduction in light passing through the
well, the optical density (OD), which is typically scaled from 0 to 4. If it is the aim
to diagnose the presence of the antigen in a sample the capture ELISA is commonly
used, i.e., the surface of the microwell is coated with a capture antibody against the
antigen. The capture of the antigen may then be directly visualized by the use of
another specific antibody conjugated with an enzyme, or indirectly by first using a
specific antibody, and then an enzyme-conjugated anti-immunoglobulin (sandwich
technique). If the aim is to demonstrate the presence of specific antibodies in a serum
sample, i.e., serology, the surface of the microwell is coated with the antigen, and the
enzyme-conjugated anti-immunoglobulin is used to demonstrate the antigen-antibody reaction. The method used for serology may also be used for demonstrating
antigen. In competitive ELISA the sample containing the antigen is mixed with a
serum sample with a known ELISA-reaction, which then is tested by ELISA serology.
A drop in the OD-value is due to antigen in the sample having used the antibodies
in the serum.

5.8). For diagnosing the presence of the antigen in a sample (Table 5.3), e.g., *Salmonella* spp. or *Listeria* spp., the capture technique is used, that is, the surface of the microwell is coated with an antibody. The application of ELISA for serology (Table 5.4), i.e., demonstration of antibodies in serum, is one of the most widely used methods in health monitoring, for example, as a tool for screening for CAR bacillus, *C. piliforme*, *Helicobacter hepaticus*, *Mycoplasma* spp., and *Pasteurella pneumotropica*. (See also Figures 5.8 through 5.11 pertaining to Tables 5.3 and 5.4.)

5.4.2   The microtiter plates

The microtiter plates are normally made of either polystyrene or PVC. The material is essential, and two types are important in bacterial serology: high binding plates, also known as enhanced binding or maxisorp, bind approximately 400 ng protein/cm^2; and medium binding plates, also known as polysorp, bind only approximately 100 ng protein/cm^2. High binding plates will — if used with a potent blocker — be the first choice, but to avoid too high a background signal it may be necessary to use medium binding plates instead. The number of wells used for bacterial ELISA is $8 \cdot 12 = 96$ (Figure 5.8). To avoid the waste of antigen one may coat single or double strips

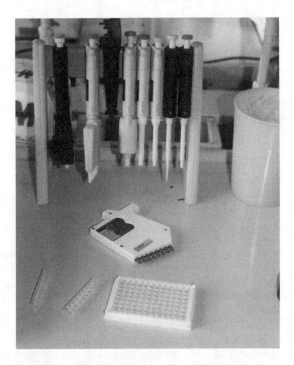

Figure 5.8   The 96-well microtiter plate, some single 8-well strips, and a multipipette for the manual application of sera and reagents.

Table 5.3 A Typical Protocol for Demonstrating the Presence of Bacteria in Samples by Enzyme-Linked Immunosorbent Assay (ELISA) (Direct Sandwich Method)

Specific antibody against the bacterium
Microtiter 8-well single strips (enhanced binding) + frame (Figure 5.8)
Parafilm
OPD tablets
10-ml glass tubes
Towel
Tinfoil
2-ml cryotubes
Positive control serum

Whirl-mixer
ELISA washer (Figure 5.9)
Multipipette with trays (Figure 5.10)

Materials

Specific antibody against the bacterium conjugated with horseradish peroxidase (HRP)
2 N Sulfuric acid
Carbonate buffer (Table 5.5)
Washing buffer (Table 5.5)
Dilution buffer (Table 5.5)
Citrate buffer (Table 5.5)
30% Hydrogen peroxide (keep cold and covered by tinfoil)
Fetal calf serum

Equipment

ELISA reader (Figure 5.11)
Magnet stirrer

Coating the wells with the antibody

The antibody is diluted 1:1000 in carbonate buffer. From the tray 0.2 ml is added to each well. The mixture is transferred to a multipipette tray. The plate with the well washer and incubated 3 min at room temperature. The plate is covered with parafilm and incubated at 4°C. After 24 h the wells are emptied by slashing the plate with the bottom upside down several times against a towel. The wells are filled with washing buffer with the plate washer and incubated 3 min at room temperature. The washing buffer is then removed and the plate is slashed against the towel. The washing procedure is repeated 6 times. All wells are then filled with the dilution buffer and incubated 2 h at room temperature (blocking); the six-step washing procedure is then repeated. After washing the wells may be stored at –20°C until use if they are filled with the washing buffer.

Preparing the OPD chromogen solution

4 tablets are grappled with tweezers (do not touch!) and placed in a small Ehrlen-Meyer bottle. 12 ml citrate buffer and a magnet are added. The solution is stirred until the tablets are fully dissolved, then 5 µl 30% hydrogen peroxide is added. The glass is packed in tinfoil and used within 30 min.

Performing the assay

1. Add 0.2 ml test sample diluted 1:100 with the dilution buffer. A blind sample (dilution buffer only) and a known positive are added to each of two wells in each plate. The position of each sample is reported on a reporting sheet or computerized. The plate is incubated at room temperature for 2 h.

2. The samples are removed with the plate washer, and the plates are washed by the six-step washing procedure as described for the coating procedure.

3. 0.02 ml HRP-conjugated antibody in a proper dilution found at pretesting is added to each well. The plate is incubated at room temperature for 2 h.

4. Repeat step 2.

5. 0.2 ml OPD chromogen-solution is added to each well (safety cabinet). The plates are incubated in the dark at room temperature for 10 min.

6. The reaction is stopped by adding 50 µl 2 N sulfuric acid.

7. The plates are read with the ELISA reader. The 492-nm filter is used for reading and the 630-nm filter is used as a reference.

Table 5.4 A Typical Protocol for Performing Bacterial Serology by Enzyme-Linked Immunosorbent Assay (Indirect Sandwich Method)

Materials

Specific antigen

Microtiter 8-well single strips (enhanced binding) + frame (Figure 5.8)

Parafilm

OPD tablets

10-ml glass tubes

Towel

Tinfoil

2-ml cryotubes

Positive control serum

Specific anti-immunoglobulin conjugated with horseradish
  peroxidase (HRP)

2 *N* Sulfuric acid

Carbonate buffer (Table 5.5)

Washing buffer (Table 5.5)

Dilution buffer (Table 5.5)

Citrate buffer (Table 5.5)

30% Hydrogen peroxide (keep cold and covered by tinfoil)

Fetal calf serum

Equipment

Whirl-mixer

ELISA washer (Figure 5.9)

Multipipette with trays (Figure 5.10)

ELISA reader (Figure 5.11)

Magnetic stirrer

Coating the wells with the antigen

160 µg of the antigen is resuspended in 0.5 ml carbonate buffer in a cryotube and whirl-mixed. The mixture is transferred to a multipipette tray to which 19.5 ml carbonate buffer is added. The suspension is mixed in the tray. From the tray 0.2 ml is added to row A, C, E, and G in a 96-well microtiter plate. To the other wells only the carbonate buffer is added. The plate is covered with parafilm and incubated at 4°C. After 24 h the wells are emptied by slashing the plate with the bottom upside down several times against a towel. The wells are filled with washing buffer with the plate washer and incubated 3 min at room temperature. The washing buffer is then removed and the plate is slashed against the towel. The washing procedure is repeated six times. All wells are then filled with the dilution buffer and incubated 2 h at room temperature (blocking): the six-step washing procedure is then repeated. After washing the wells may be stored at –20°C until use if they are filled with the washing buffer.

Preparing the OPD chromogen solution

4 tablets are grappled with tweezers (do not touch!) and placed in a small Ehrlen-Meyer bottle. 12 ml citrate buffer and a magnet are added. The solution is stirred until fully dissolved, then 5 µl 30% hydrogen peroxide is added. The glass is packed in tinfoil and used within 30 min.

Performing the assay

1. Add 0.2 ml test serum diluted 1:100 with the dilution buffer. Each test serum is added to an antigen-coated well and a control well (Figure 5.12). Positive and negative controlsera are added in the same way in every plate. The position of each serum is reported on a reporting sheet or computerized. The plate is incubated at room temperature for 2 h.
2. The sera are removed with the plate washer, and the plates are washed by the six-step washing procedure as described for the coating procedure.
3. 0.02 ml species-specific, HRP-conjugated anti-immunoglobulin in a proper dilution found at pretesting is added to each well. The plate is incubated at room temperature for 2 h.
4. Repeat step 2.
5. 0.2 ml OPD chromogen solution is added to each well (safety cabinet). The plates are incubated in the dark at room temperature for 10 min.
6. The reaction is stopped by adding 50 µl 2 N sulfuric acid.
7. The plates are read in the ELISA reader. The 492-nm filter is used for reading and the 630-nm filter is used as a reference.

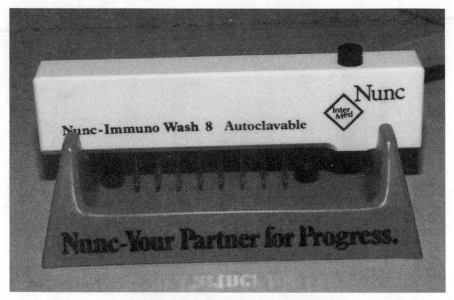

Figure 5.9   A simple, manual microplate washer.

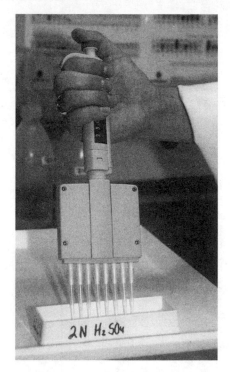

Figure 5.10   Multipipette with trays.

Figure 5.11 ELISA reader.

with 8 or 16 wells. Strips with different antigens can be collected in the same frame to make up a 96-well ELISA plate. The volume of the wells may vary from 50 to 250 µl. The higher the working volume the lower the detection limit. Precoated plates from Charles River (U.S.) are 250 µl but only intended for a working volume of 50 µl. This author prefers working with 200 µl in 250-µl wells.

5.4.3   The antigen

In principle, bacteria may be used directly as antigen, i.e., they are suspended in the coating buffer (see below) and the well is coated with the buffer suspension. In most assays it is advantageous to prepare the antigen for coating, minimally to prevent continous growth. The simplest way is boiling the antigen for 75 min. Smaller particles may be achieved by sonication, and proteins may be extracted by the use of certain chemicals.

5.4.4   Antibodies, enzymes, and substrates

Enzyme-conjugated, species-specific anti-immunoglobulins are available from several commercial suppliers, e.g., Sigma (U.S.), DAKO Corporation (U.S.), Charles River (U.S.), Harlan (U.K.) and ICN (U.S.). They should usually be diluted in the range from 1:1000 to 1:4000. If this has not been determined by producer pretesting this has to be done in the laboratory prior to use. Anti-immunoglobulins for some species may be difficult to find at commercial suppliers. This problem may be solved by testing an anti-immunoglobulin to a related species to identify if it binds acceptably to the immunoglobulin of the missing species; for example, this author has successfully used Cappel® antihamster (ICN, U.S.) when testing gerbils.

The most common conjugate enzyme is horseradish peroxidase, but alkaline phosphatase, β-galactosidase, and urease also may be applied. The substrate is, of course, specific to the enzyme. Furthermore, the substrate for ELISA has to be soluble. Table 5.6 lists enzymes, relevant substrates, and the chemical reaction they catalyze. Substrates are supplied by several commercial suppliers, e.g., Sigma (U.S.) and DAKO Corporation (U.S.).

5.4.5   Coating the wells

Although absorption of protein to the plastic surface of the well is based on hydrophobic binding and thereby relatively independent of pH, a buffer with pH as or slightly above the pH_{ISO} is generally used, the most common one being a pH 9.6 carbonate buffer (Table 5.5). The buffer should not contain any detergents because they will block the binding sites and thereby reduce the amount of antigen bound. The coating time should not be too short. Either 4 h at 37°C or 24 h at 4°C is generally usable. Care should be taken to ensure the same temperature in each well by placing the plates in a closed box, e.g., vaporization is avoided by covering the plates with parafilm. The coating time may be reduced if the plates are shaken on a plate-shaker during coating.

To eliminate the impact of unspecific binding on the interpretation of the result it is an option to coat rows with both antigen and pure buffer within the same plate (Figure 5.12).

5.4.6   Blocking the wells

Adding Tween 20 to the dilution buffer is necessary to avoid unspecific binding on nonantigen-covered plastic. Tween 20 alone may, however, be insufficient, and therefore some kind of nonimmunoglobulin binding protein is generally used as a blocking agent. The blocking agents are added to the dilution buffer in a concentration typically of 5%. The most common blocking agent is bovine serum albumin (BSA), but various types of serum are also used. To avoid specific antibodies fetal calf serum is preferable when using serum for blocking. Alternatively, cow milk is a cheap but reliable option. A blocking step, in which the wells are filled with the dilution buffer with 5% blocking agent, should be introduced in the protocol after coating but prior to testing the samples.

Table 5.5 Examples of Recipes for Buffers used for ELISA

Carbonate buffer		Washing buffer	
Na_2CO_3, $10H_2O$	4.29 g	Tween 20	1 ml
$NaHCO_3$,	2.93 g	NaCl	15 g
Distilled water ad	1000 ml	PBS ad	1000 ml

Dissolve the carbonates in 800 ml water. Adjust pH to 9.6 with $1M$ HCl or NaOH. Dilute ad 1000 ml. Store at 4°C.

Dissolve the Tween and salt in 800 ml water. Adjust pH to 7.8 with $1M$ HCl or NaOH. Dilute ad 1000 ml. Store at 4°C.

Dilution and blocking buffer		Citrate buffer	
Fetal calf serum	10 ml	Citric acid, H_2O	7.30 g
Tween 20	0.2 ml	Na_2HPO_4, $12H_2O$	23.88 g
PBS ad	200 ml	Distilled water ad	1000 ml

Dissolve the Tween and the serum in 800 ml water. Adjust pH to 7.2 with $1M$ HCl or NaOH. Dilute ad 1000 ml. Should be used immediately or stored at –20°C.

Dissolve the acid and phosphates in 800 ml water. Adjust pH to 5.0 with $1M$ HCl or NaOH. Dilute ad 1000 ml. Store at 4°C.

5.4.7 Performing the assay

A typical protocol for capture ELISA is shown in Table 5.3 and for ELISA serology in Table 5.4. The method must be optimized individually from laboratory to laboratory and from test to test. Having optimized all the above factors, such as the type of plates, the antigen, etc., the temperatures and times of incubation still have to be considered. Incubation temperature may range from 4° to 40°C, but in practice there is a choice between 4°C (fridge), 20°C (room temperature), and 37°C (incubator). The highest achievable enzymatic activity is not always optimal, as this also leads to more nonspecific reactions. Likewise, the incubation time has to be considered. The longer the time, the more antigen-antibody reactions, but also the more unspecific reactions. At a certain level no more antigen-antibody reactions will occur as a result of increased temperature or incubation time, and, in general, a raise in temperature means shorter incubation time. For example, the same result might be achieved at 4°C for 24 h, at 20°C for 2 h, and at 37°C for 30 min. Also, shaking the plates on a plate shaker enhances the antigen-antibody reactions and might thereby reduce the incubation time.

5.4.8 Control sera

Control sera may derive either from spontaneously infected animals or from immunized animals. Sera from germ-free animals are directly usable as

Table 5.6 Enzymes Used for Conjugation with Antibodies in ELISA and the Related Substrates, the Reactions Catalyzed, and the Color/Wavelength Produced during the Reaction

Enzyme		Substrate	Color/wavelength	Chemical reaction
Horseradish peroxidase	TMB	Tetramethyl-benzidine hydrochloride	Yellow/492 nm	2 Substrate-H + $H_2O_2 \rightarrow$ 2 Substrate + 2 H_2O
	OPD	O-Phenylenediamine		
	ABTS	2,2′-Azino-di(3-ethyl-benzthizoline) sulfonate-6-diammonium salt		
Alkaline phosphatase	PNPP	P-Nitrophenylphosphate	Yellow/492 nm	PNPP $\rightarrow$ phosphate + P-nitrophenol
β-Galactosidase	ONPG	O-Nitro-phenylgalactoside	Yellow/492 nm	ONPG $\rightarrow$ galactose + O-nitrophenol

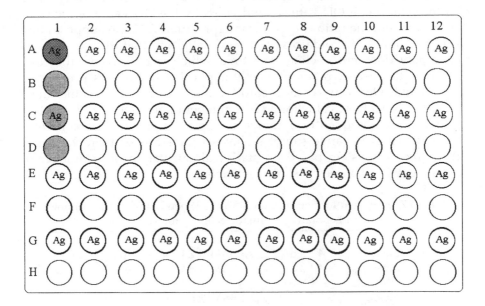

Figure 5.12 A system for setting up a 96-well microtiter plate for ELISA. Only half of the rows (A, C, E, and G) are coated with the antigen, while the remaining rows (B, D, F, and H) during the coating step are filled with the pure coating buffer only. A1 and B1 are filled during the test with a postive control serum, and C1 and D1 are filled with a negative control serum. Test sera are filled into each coated and non-coated well. A high OD-value in A1 is expected, while B1, C1, and D1, as well as all wells in rows B, D, F, and H should have OD-values far below the cutoff value.

negative controls. Positive control sera should be pretested in different dilutions, e.g., 1:25, 1:50, 1:100, and 1:200, and a dilution giving an OD-value within an acceptable and usable range, e.g., between 1 and 2.5, should be chosen as standard for that serum. Commercially available control sera may be diluted according to instructions given by the supplier, but pretesting at lower dilutions in the laboratory is recommended. For all dilutions, the dilution buffer (Table 5.5) is used. Control sera may be stored at –20°C, but they must be stored in small volumes since they should be thawed only once.

5.4.9   Interpretation of the OD-value

As the main aim of routine health monitoring of laboratory animals is to diagnose the infection on colony and not on an individual level, it is highly recommended to use high cutoff values for bacterial serology and counteract the drop in sensitivity by sampling more animals. If noncoated wells are included for every sample, the value used for interpretation of a certain serum should be the difference between the OD-value of the coated and noncoated well inoculated with that serum. To eliminate day-to-day variation some laboratories include a standard serum in the assay. Prior to interpretation all OD-values are divided by the OD-value of the standard serum and the ratio is used for interpretation. This presupposes that a certain amount of standard serum is available and that the OD-value of this supply does not decline over time, which is not always the case. Therefore, the use of ratios is mostly applicable in projects run over a short period. One way to set a cutoff value is to choose a value much lower than the lowest value found in a number of positive samples tested, e.g., 0.2, and subsequently optimize the assay to ensure that none of a high number of negative samples shows values higher than the cutoff value. Another system based on statistical calculation on a number of known negative samples is shown in Table 5.7.[5]

Table 5.7 A System for Setting Cutoff Values When Using ELISA Serology
in Laboratory Animal Health Monitoring

Negative	$mean_{negatives} + 4 * s.d._{negatives} > $ SAMPLE OD	$p \geq 0.0313$
+	$mean_{negatives} + 4 * s.d._{negatives} \leq $ SAMPLE OD $> mean_{negatives} + 8 * s.d._{negatives}$	$0.0313 > p \geq 0.0078$
++	$mean_{negatives} + 8 * s.d._{negatives} \leq $ SAMPLE OD $> mean_{negatives} + 12 * s.d._{negatives}$	$0.0078 > p \geq 0.0035$
+++	$mean_{negatives} + 12 * s.d._{negatives} \leq $ SAMPLE OD	$0.0035 > p$

Note: The mean and the standard deviation of the OD values found in a number of samples from noninfected colonies are used for estimation of a cutoff value according to Chebyshev's theorem. The p-value describes the highest possible probability that a sample OD value within a given range should belong to the same normal distribution as the negative samples pretested.

5.5   *Immunoblotting*

Antibodies may be detected by so-called immunoblotting.[6] In *Western blotting* proteins of the bacterium are separated by sodiumdodecyle sulfate electrophoresis (SDS-PAGE) and are thereafter transferred to an immobilizing membrane, in which they are irreversibly bound. Transfer may be done in different ways, e.g., by the buffer tank blotting technique (Figure 5.13). If the proteins are placed directly upon the membrane the technique is called *dot blotting* instead of Western blotting. The membrane is made of nitrocellulose, or alternatively, polyvinylidene chloride or nylon if the chemiluminescent detection technique is to be applied. After blocking free protein sites the test serum is added and antigen-antibody reactions with the proteins are visualized by using fluorescein-, enzyme-, or radionucleotide-labeled anti-immunoglobulin.

In laboratory animal bacterial serology the method has not been as widely used as IFA and ELISA.

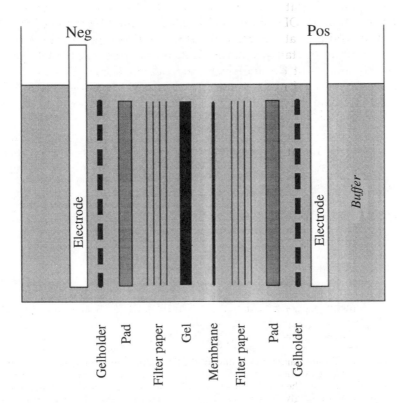

Figure 5.13 The buffer tank blotting technique. Proteins are transferred from an electrophoretic gel to an immobilizing membrane, in which they are irreversibly bound.

References

1. Needham, J., personal communication, 1997.
2. Kraft, V. and Meyer, B., Diagnosis of murine infections in relation to test methods employed, *Lab. Anim. Sci.*, 36, 271, 1986.
3. Hansen, A. K. and Skovgaard Jensen, H. J., Experiences from sentinel health monitoring in units containing rats and mice in experiments, *Scand. J. Lab. Anim. Sci.*, 22(1), 1, 1995.
4. Naish, S. J., Boenisch, T., Farmilo, A. J., and Stead, R. H., Immunochemical Staining Methods, DAKO Corporation, Carpinteria, 1989.
5. Hansen, A. K., Thomsen, P., and Skovgaard Jensen, H. J., A serological indication of the existence of a guinea pig poliovirus, *Lab. Anim.*, 31(3), 212, 1997.
6. Bjerrum, O. J. and Heegård, N. H. H., *Handbook of Immunoblotting of Proteins*, CRC Press, Boca Raton, FL, 1988.

chapter six

Molecular biological methods

Contents

6.1   Detection by molecular probes...105
      6.1.1   Solid-phase hybridization. 105
      6.1.2   Solution-phase hybridization 105
6.2   Amplification of nucleic acids..105
      6.2.1   Principle 105
      6.2.2   The polymerase enzyme...108
      6.2.3   Reverse transcriptase RNA PCR..108
      6.2.4   The template sequence and the primers 109
      6.2.5   Preparation of the sample...109
      6.2.6   Detection of the PCR product 109
      6.2.7   Contamination of the amplification process..............................110
6.3   Restriction analysis of chromosomal DNA...110
References..111

Molecular biological methods are generally divided into two types: those in which the DNA — occasionally RNA — is already present in detectable amounts, and those in which the DNA or RNA must be amplified before detection.

To be able to apply molecular biological methods for detection or identification of an infectious agent one or several specific sequences from that organism must be known. Tables 6.1 and 6.2 provide examples of known sequences for some laboratory animal bacteria. Further relevant specific sequences may be found in the GenBank Database run by the National Center for Biotechnology Information, which can be reached at http://www.ncbi.nlm.nih.gov/irx/genbank/query_form.html. Some known sequences are short, e.g., 400 to 600 base pairs, while others are far longer, e.g., 1400 to 2000 base pairs.

Table 6.1 16S Ribosomal RNA Gene Sequences of the Loci CLOPILIFRG, S82325, and S72349[10] of *Clostridium piliforme*

Locus	Pos								
CLOPILIFRG	1	atgaacgctg	gcggcgtgcc	taacacatgc	aagtcgagcg	gagatattat	agcgcgtgcg	tcangatatt	ttagcggcgg
	81	acngngagt	aacgcgnggg	taacctgccc	tatacactgg	gataacatcg	agaaatcggt	gctaatacca	gataagctaa
	161	cagtaaggca	tcttacagtt	agaaaaactg	aggtggtata	ggaggggccc	gcgtctgatt	agctagttgg	tgtggtaaaa
	241	gcataccaag	gcaacgatcn	gtagccgacc	tgagagggtg	atcggccacn	ttggaactga	gacacggtcc	aactcctacg
	321	ggaggcagca	gtggggaata	tngcacaatg	gggggaaccc	tgatgcagca	acgccgcgtg	aaggaagaag	tatttcggta
	401	tgtaaactc	tatcgacagg	gaagaaaaaa	atgacggtac	ctgaatnaga	agcaccggct	aaatacgtgc	cagcagccgc
	481	ggtaatacgt	atggngcaag	cgttatccgg	attacncggg	ngtaaagggn	gagtaggcgg	ttagataagt	catangr gaa
	561	atttctgggc	tcaactccag	cgcngcataa	gaaactattt	aactagagta	caggagaggt	aagcggaatt	cctagtgtag
	641	cggtgaaatg	cgtagatatt	aggaagaacn	ccggtggcga	aggcggctta	ctggactgaa	actgacgctg	agtcacgaaa
	721	gcgtggggag	cgaacaggat	tagataccct	ggtagtccac	gctgtaaacn	atgagtgcta	ggtgttgggg	agaaattctc
	801	ggtgccgag	caaacgcaat	aagcactcca	cctggggagt	acgacngcaa	ggttgaaact	caaaggaatt	gacggggacc
	881	cgcacaggt	ggagcatgtg	gtttaattcg	aagcaaacgcg	aagaacctta	cctaaacttg	acataccatt	gacagac ac
	961	gtaaagtagt	tttccttcgg	gacaatggat	catggttgtc	catggttgtc	gtcagctcgt	gtcgtgagat	gttgggttaa
	1041	gtcccgcaac	gagcgccacc	cctattctta	gtagccagca	cctggggtgg	gcactctagg	gagactgccg	tggataacmn
	1121	ggaggaaggt	ggggatgacg	tcaantcatc	atgccccta	tgtttagggc	tacacacgtg	ctacaatggc	tacaacaaag
	1201	tgaagcgaga	cagtaatgtg	gagcaaagca	cataaaagta	gtcccagttc	ggattgtagt	ctgcaactg	actacatgaa
	1281	gttggaatcg	ctagtaatcg	cgaatcagaa	tgtcgcggtg	aatacgttcc	cgggtcttgt	acacaccgcc	cgtcnacgat
	1361	gggagttgga	agcgcccgaa	gcctgtgacc	taaccgcaag	ggaggagcag	tcgaaggtga	agccagtgac	tggggtgaag
	1441	tcgtaacaag	gtatc						
S82325	1	aattccgagg	aaggtgggga	tgacgtcaaa	tcatcatgcc	ccttatgttt	agggctacac	acgtgctaca	atggctacaa
	81	caaagtgaag	cgagacagtg	atgtggagca	aagcacataa	aagtagtccc	agttcggatt	gtagtctgca	actcgactac
	161	atgaagttgg	aatcgctagt	aatcgctagt	cagaatgtcg	cggtgaatac	gttcccgggc	ctgggatccc	
S72349	1	aacaggatta	gatacctgg	tagtccacgc	tgtaaacgat	gagtgctagg	tgttgggaag	aaattctcgg	g
	81	aacgcaataa	gcactccacc	tggggagtac	gaccgcaagg	ttgaaactca	aaggaattga	cggggacccg	tgcccagca
	161	ggagcatgtg	gtttaattcg	aagcaacgcg	aagaacctta	cctaaacttg	acataccatt	gacaggctac	cacaagcggt gtaaagtag

241	tttccttcgg	ggcaatggat	acaggtggtg	catggttgtc	gtcagctcgt	gtcgtgagat	gttgggttaa	gtccgcaac
321	gagcgcaacc	ccatttctta	gtagccagca	cttgggtgg	gcactctaag	gagactgccg	tggataacac	ggaggaaggt
401	gggatgacg	tcaaatcatc	atgccccta	tgtttagggc	tacacacgtg	ctacaatggc	tataacaaag	tgaagcggaga
481	cagtgatgtg	gagcaaagca	cataaaagta	gtcccagttc	ggattgtagt	ctgcaactcg	actacatgaa	gttggaatcg
561	ctagtaatcg	cgaatcagaa	tgtcgcggtg	aatacgttcc	cgggtcttgt	acacaccgcc	cgtca	

Table 6.2 Complete Sequences of the 16S Ribosomal RNA Loci PASRRNAA[11] and AF012090[12] of *Pasteurella pneumotropica*

AF012090								
1	ataaacgcgg	gcggcattct	taacacatgc	aagtcgaacg	gtagcaggaa	ggaagcttgc	tttctttgct	gacgagtggc
81	ggacggggtga	gtaatgcttg	ggaatctggc	ttatggaggg	ggataactgc	gggaaactgc	agctaatacc	gcgtaaagtc
161	tttggactaa	aggggggcgtt	tgctcttgcc	ataagatgag	cccaagtggg	attaggtagt	tggtggggta	atggctcacc
241	aagccgtcga	tctctagctg	gtctgagagg	atgaccagcc	acacgggac	tgagacacgg	cccggactcc	tacgggaggc
321	agcagtgggg	aatattgcgc	aatgggggga	acctgacgc	agccatgccg	cgtgaatgaa	gaaggccttc	gggttgtaaa
401	gttctttcgg	tgatgaggaa	ggcagtttgg	ttaatagcca	agctgattga	cgttagtcac	agaagaagca	ccggctaact
481	ccgtgccagc	agccgcggta	atacggaggg	tgcgagcgtt	aatcggaata	actgggcgta	aagggcacgc	aggcggatttt
561	ttaagtgagg	tgtgaaagcc	ccgggcttaa	cctgggaatt	gcatttcaga	ctgggaatct	agagtacttt	agggaggggggt
641	agaattccac	gtgtagcggt	gaaatgcgta	gagatgtgga	ggaataccga	aggcgaaggc	agccccttgg	gaatgtactg
721	acgctcatgt	gcgaaagcgt	gggagctaaa	caggattaga	taccctggta	tacccgctgt	aaacgctgt	taaaactcaa
801	gttggggttt	ggggttaagt	cccgtagcta	acgtgataaa	tcgaccgtct	ggggagtacg	gccgcaaggt	tactcttgac
881	tgaattgacg	ggggccccgca	caagcggtgg	agcatgtggt	ttaattcgat	gcaacgcgaa	gaacctaacc	cagtctcgt
961	atccagagaa	tcctgtagag	atacgggagt	gcctaggga	gctttgagac	agtgctgca	tggctcgt	actaaagga
1041	tgtgaaagt	tgggttaagt	cccgcaacga	gcgcaacct	tatcctttgt	tgccagcgat	acggtcggga	cacacgtgct
1121	gactgccggt	gataaaccgg	aggaaggtgg	ggatgacgtc	aagtcatcat	ggcccttacg	agtaggggcta	attggagtct
1201	acaatggcgt	atacagaggg	aggcgaagca	gcgatgtgga	gcgaatctca	caaagtacgt	ctaagtccgg	ggccttgtac
1281	gcaactcgac	tccatgaagt	cggaatcgct	agtaatgcgg	aatcagaatg	tcgcggtgaa	tacgttccg	ccacgtatg
1361	acaccgcccg	tcacaccatg	ggagtgggtt	gtaccagaag	tagatagctt	aaccgtaagg	ggggcgttta	ccacgtatg
1441	attcatgact	ggggtgggaat	cgt					

continued

Table 6.2 (continued) Complete Sequences of the 16S Ribosomal RNA Loci PASRRNAA[11] and AF012090[12] of *Pasteurella pneumotropica*

PASRRNNAA								
1	aattgaagag	tttgatcang	gctcagattg	aacgctggcg	gcaggcttaa	cacatgcaag	tcgaacggta	acaggagaaa
81	gcttgctntc	ttgctgacga	gtggcggacg	ggtgagtaat	gcttgggaat	ctggcttatg	gagggggata	actactggaa
161	acggtagcta	ataccgcata	aggtctaagg	acaaaagggg	gcgtaagctc	ttgccataag	atgagcccaa	gtgggattag
241	gtagttggtg	aggtaatggc	tcaccaagcc	tncgatctct	agctggtctg	agaggatgac	cagccacacc	gggactgaga
321	cacggcccng	nctcctacgg	gaggcagcag	tggggaatat	tgcgcaatgg	ggggaaccct	gacgcagcca	tgccgcgtga
401	atgaagaagg	ccttcgggtt	gtaaagttct	ttcggtgatg	aggaaggtga	taaggttaat	acccttatta	attgacgtta
481	gtcacagaag	aagcaccggc	taactccgtg	ccagcagccg	cggtaatacg	gagggtgcga	gcgttaatcg	gaataactgg
561	gcgtaaagggg	cacgcaggcg	gattttaag	tgaggtgtga	aagccccggg	cttaacctgg	gaattgcatt	tcagactggg
641	aatctagagt	actttaggga	gggntagaat	tccacgtgta	gcggtgaaat	gcgtagagat	gtggaggaat	accgaaggcg
721	aaggcagccc	cttgggaatg	tactgacgct	catgtgcgaa	agcgtggggga	gcaaacagga	ttagataccc	tggtagtcca
801	cgctgtaaac	gctgtcgatt	tggggggttgg	gctttaagct	tggcgccgt	agctaacgtg	ataaatcgac	cgcctgggga
881	gtacggccgc	aaggttaaaa	ctcaaatgaa	ttgacggggg	cccgcacaag	cggtggagca	tgtgtttaa	ttcgatgnaa
961	cgcgaagaac	cttacctact	cttgacatcc	agagaagaga	ctagagatag	ttttgtgcct	tagggaactt	tgagacaggt
1041	gctgcatggc	tgtcgtcagc	tcgtgttgtg	aaatgttggg	ttaagtcccg	caacgagcgc	aaccttatc	ctttgttgcc
1121	agcgacttgg	tcgggaactc	aaaggagact	aaccgggaga	aaccgggaggga	aggtggggat	gacgtcaagt	catcatggcc
1201	cttacgagta	gggctacaca	cgtgctacaa	gccggtgata	agagggaagc	gaagcagcga	tgtggagcga	atctcataaa
1281	gtgcgtctaa	gtccggattg	gagtcgtcaa	tggcgtatac	tgaagtcgga	atcgctagta	atcgcgaatc	agaatgtcgc
1361	ggtgaatacg	ttcccgggcc	ttgtacacac	ctcgactcca	accrtgggag	tgggttgtac	cagaagtgga	tagcttaacc
1441	gaaagggggggg	cgttcaccac	ggtatgattc	cgccgtcac				
	a							

6.1   Detection by molecular probes

A probe is an oligonucleotide able to anneal to a defined sequence in the
microbial DNA. Such are commercially available, e.g., from Gen-Probe Inc.
(U.S.). Of interest in laboratory animal bacteriology are probes for direct
detection of group A *Streptococcus* and for the identification of cultures of
Streptococcus, Staphylococcus, Mycobacterium, and *Listeria.* The probes may be
radio- or enzyme-labeled to help detection.

6.1.1   Solid-phase hybridization

In the *dot blot* bacterial cells are lysed; the DNA is denaturized and fixed
onto a nylon membrane, which is inserted into a solution containing the
probes. After a proper reaction time, the unbound probe is washed away
from the membrane.

   In situ hybridization is in principle the same, but the DNA is present in
tissue that has been sliced and embedded in paraffin, thereby allowing the
detection of specific DNA in the tissue in order to facilitate a causal diagnosis.
Disadvantages are a low sensitivity as well as difficulties in preserving the
tissue while denaturating the DNA.

6.1.2   Solution-phase hybridization

In *solution-phase hybridization* the target DNA and the probe are both non-
bound in an aqueous solution. The probe needs to be a single-stranded DNA,
which does not hybridize with itself. When the probe has hybridized with
the target DNA the remaining single-stranded DNA is digested by an S2
nuclease and the double-stranded DNA may be precipitated by trichloro-
acetic acid or bound to a hydroxyapatite column.

6.2   Amplification of nucleic acids

6.2.1   Principle

In the *polymerase chain reaction (PCR)*[1] (Figures 6.1 and 6.2) small amounts of
DNA are amplified by a factor up to 10^6. The principle is rather simple: at
91° to 94°C the double-stranded DNA is denatured into two single strands,
at 52°C two oligonucleotide primers will anneal to their respective recogni-
tion sites, and at 72° to 80°C a polymerase will catalyze resynthesis from the
one end of the primers. This cycle is repeated for a number of times in a so-
called thermocycler and, finally, a certain amount of specific DNA has been
produced. A procedure including n cycles should in principle result in 2^n
times as much DNA, but as the efficiency rate of each cycle is not 100% but
less, it is expected that the amount of DNA is less, i.e., (1 + efficiency rate
per cycle)^n times as much DNA will be produced.

   The amount of the various reagents, the temperatures, and the duration
of each step in the cycle have to be optimized for each single assay, i.e., the

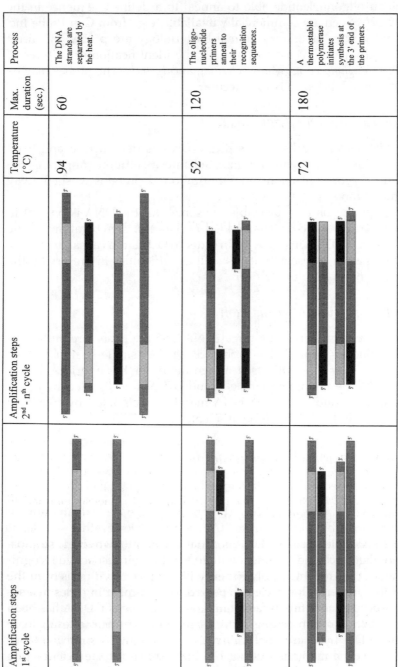

Figure 6.1 Principles of a polymerase chain reaction (PCR) I. The microbial DNA in the sample is propagated in a number of cycles. In the first cycle the product may be somewhat longer than the 3' end of the reconstructed primer annealing sequence, but during the next cycle, transcription of the first product will result in a product including the two primer annealing sequences as well as the sequence between them, referred to as *the short product*. As all products transcribed from the short products also will be short products, only those transcribed directly from the template DNA will be somewhat longer, and the final product will consist mostly of short products. This means that if electrophoresis is used for detection the short product will make one broad band.

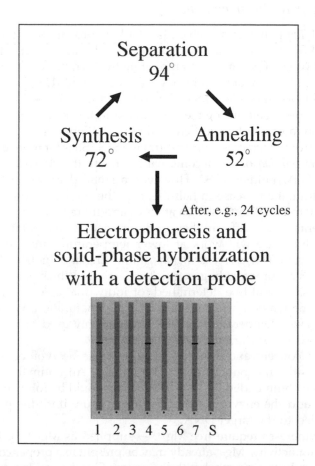

Figure 6.2 *Principles of a polymerase chain reaction (PCR) II.* After a number of cycles in the thermocycler according to the principles shown in Figure 6.2, the molecular weight and the sequence of the product may be subjected to electrophoresis, transferred to a membrane, and detected with a specific probe. In this example, out of seven samples tested against a standard of the short product, Nos. 1, 4, 7, and, obviously, the standard (S), are positive. Other means of detection may be applied as well.

temperatures and durations given in Figures 6.1 and 6.2 are only approximate proposals.

Obviously, the DNA or RNA will only be propagated if present in the initial sample, and as such the method is excellent for detecting the presence of infectious agents in samples from animals. That means the method may be the one method for infections, which are persistent in the animal but difficult to isolate by traditional cultivation techniques, e.g., *Helicobacter hepaticus,*[2] while the method is of a more limited use for the examination for nonpersisting infections.

6.2.2   The polymerase enzyme

The original *Taq* polymerase, which is a 94-kDa protein of 832 amino acids, was isolated from the thermophilic bacterium *Thermus aquaticus*. Today, however, several types of polymerases originating from a range of different bacteria are commercially available from, e.g., Perkin-Elmer (U.S.), New England Biolabs (U.S.), or Stratagene (U.S.). The polymerase has to be thermostable not to be denatured itself during the DNA denaturation step at 91° to 94°C, but different polymerases show different thermostability; e.g., at 95°C the *Pfu* polymerase from *Pyrococcus furiosus* (Stratagene, U.S.) is far more thermostable than the original *Taq* or the recombinant version, the *AmpliTaq* polymerase (both from Perkin-Elmer, U.S.). However, a small decrease in temperature leads to a dramatic increase in half-life; e.g., the *AmpliTaq* polymerase has a half-life of 10 min at 97.5°C, 40 min at 95°C, and more than 130 min at 92.5°C. The polymerases also differ in their optimum activity temperatures; e.g., the *AmpliTaq* polymerase will work at lower temperatures than, e.g., the *Vent* polymerase from *Thermococcus literalis* (New England Biolabs, U.S.). Some polymerases have proofreading exonuclease activity. Which polymerase to use for which assay should be determined for individual cases, judged, e.g., on the basis of the temperature needed for DNA denaturation. When using the *Taq* polymerase a temperature of 72°C is commonly used to reduce primer-template dissociation during the extension phase.[3]

Too high concentrations of the enzyme should be avoided, especially if an enzyme with an exonuclease activity is used. An optimal concentration generally is recommended by the vendor and should be followed. It may be wise not to add the enzyme to the reaction mixture until the primers have been annealed to the target DNA.

All polymerases require free Mg^{++} — supplied as $MgCl_2$ or $MgSO_4$, for example — for activity. Mg^{++} already may be present in a proper concentration in the reaction buffer supplied with the enzyme by the vendor. Alternatively, a Mg^{++}-free buffer may be purchased with a Mg^{++} solution enabling optimization of the Mg^{++} concentration in each assay. At higher Mg^{++} concentrations there is an increased risk of *mismatching*, i.e., the primers anneal to nonspecific sequences leading to a reduced specificity. A concentration of 1 to 4 mM is optimal for many assays, but some polymerases are more active at higher concentrations. Also, the addition of supplementary reagents, e.g., dimethylsulfoxide[4] or glycerol,[5] may enhance the sensitivity. The pH optimum for amplification is usually 7.0 to 7.5. The four free nucleotides — dATP, dCTP, dGTP, and dTTP — must be added in excessive amounts, i.e., a concentration of 40 to 200 µM of each, to avoid the reaction running out of "building material."

6.2.3   Reverse transcriptase RNA PCR

Bacteria contain thousands of copies of specific rRNA, and therefore amplification of these is often more sensitive than amplification of bacterial DNA. The recombinant *rTth* DNA polymerase from *Thermus thermophilus* (Perkin-

Elmer, U.S.) is a 94-kDa thermostable recombinant DNA polymerase, suitable for amplification of RNA targets. This polymerase is able to reverse transcribe RNA to cDNA in the presence of Mn^{++} ions, as well as to act as a DNA polymerase for PCR amplification. High temperatures (60° to 70°C) reverse transcription with *rTth* DNA polymerase and allow for efficient cDNA synthesis from RNA templates that contain complex secondary structure. PCR amplification can be performed either after chelation of the Mn^{++} and subsequent addition of $MgCl_2$ or in a single buffer system. This technique is referred to as reverse transcriptase RNA PCR, RT-RNA PCR, or simply RT-PCR.[6]

6.2.4 The template sequence and the primers

The selected region to be amplified, the *template*, normally has a size of 100 to 400 base pairs, while primers of a length of 20 to 23 base pairs often are optimal. The template sequence must be specific (highly conserved) for that organism, e.g., *Pasteurella pneumotropica*, or for a larger group searched for, e.g., Pasteurellaceae. It is also essential to select primer sequences in a way that the primers annealing to themselves or to one another is avoided. To reduce the risk of mismatching, the G + C content of the primers should be higher than the template G + C content, as guanosine-cytosine forms three hydrogen bonds compared to the two bonds formed by adenosine-thymidine.

The annealing temperature has to be optimized individually for each assay. Too high a temperature reduces the chances of annealing, while too low a temperature enhances the risk of mismatching. It is recommended to start with new primers at a low temperature, e.g., 35°C, and increase the temperature stepwise until the optimum has been found.

The two primers must be added in excess to avoid the reaction from premature cessation.

6.2.5 Preparation of the sample

If identification of a pure culture is the sole aim of PCR, the task is obviously simpler than if the aim is the detection of small amounts of the agent in samples of tissue, blood, urine, or other materials. The simplest means of sample preparation is boiling the sample in sterile water and subsequently using the boiled preparation directly in the PCR. For Gram-negative bacteria a lysis buffer may be used, e.g., a buffer containing 1% Triton X-100, 20 m*M* Tris, and 2 mM EDTA (pH 8.5).[3] Unpurified samples contain a number of uncharacterized polymerase inhibitors, which have to be removed by the use of proteinases, extraction with phenol-chloroform, or binding the DNA to silica to optimize the assay.

6.2.6 Detection of the PCR product

Ethidium bromide produces fluorescence with double-stranded DNA, so the DNA products may be visualized by simply adding ethidium bromide to the amplification reaction. This may be sufficient if the sample does not contain essential amounts of other types of DNA, as serum samples do.

Such ethidium bromide staining in the reaction tube may not be sufficient as the sole method of detection and, therefore, the PCR product may be placed in an agarose gel and subjected to electrophoresis. The simplest way of visualizing the presence of DNA in the gel is by ethidium bromide staining. However, it is preferable to use a probe for detection, as this will only detect DNA with the specific sequence. The DNA in the gel should therefore be transferred to a nitrocellulose or nylon membrane, on which the DNA is hybridized to a probe. Such probes were previously labeled with ^{32}P, ^{35}S, or ^{125}I, and detection was performed by placing the membrane on an X-ray film. Nowadays, it is more common to use a nonradioactive labeling, e.g., dioxigenin[7] or horseradish peroxidase.[8]

A simple enzymatic assay system using reverse transcription polymerase chain reaction (RT-PCR) products in a microtitration plate has been used for detection of *Clostridium piliforme*. This RT-PCR is performed with a biotin-labeled primer. The amplified cDNA is hybridized to an alkaline phosphatase-labeled DNA probe in a microtube, and the mixture is applied to a microtitration well precoated with streptavidin. Hybridization signals in the microtitration plate are visualized by reaction with substrate for alkaline phosphatase.[9]

6.2.7   Contamination of the amplification process

It is a sound principle to separate sample preparation and amplification in two separate rooms to avoid contamination with nonsample DNA during the preparation process. The amplified PCR products may, if not kept safely in the tubes, contaminate all future assays and thereby reduce their specificity. Having said this, it should also be mentioned that this problem is often given more attention than really justified.

6.3   Restriction analysis of chromosomal DNA

Some bacteria, e.g., Mycoplasmas, are not easily identified by traditional biochemical methods. Restriction endonucleases will cut the bacterial DNA at each site where a specific sequence is recognized. A six- or eight-base-pair sequence recognizing an enzyme typically may be used to cut the bacterial DNA into hundreds of fragments in lengths ranging up to 50 kb. The digested DNA is separated by constant-field agarose gel electrophoresis, which afterward is stained with ethidium bromide and viewed under UV illumination. A standard of the agent to be identified is digested along with the test isolate and the profiles are compared. Confusion may arise if the test isolate or the standard culture contains plasmides, which falsely may add to the bacterial profile. Restriction endonucleases are commercially available from New England Biolabs (U.S.).

References

1. Mullis, K. B., The unusual origin of the polymerase chain reaction, *Sci. Amer.*, 262, 56, 1983.
2. Charles River Laboratories, Helicobacter Infection in Laboratory Mice: History, Significance, Detection and Management, http://www.criver.com/techdocs/helico-1.html, 1995.
3. Cox, P. T., Don, R. H., and Mattick, J. S., Rapid animal disease diagnosis: the future is PCR, *Lab Anim. (U.S.A.)*, 2(5), 22, 1991.
4. Hung, T., Mak, K., and Fong, K., A specific enhancer for polymerase chain reaction, *Nucleic Acids Res.*, 18, 4953, 1990.
5. Dermer, S.J. and Johnson, E.M., Rapid DNA analysis of α_1-antitrypsin-deficiency: application of an improved method for amplifying mutated gene sequences, *Lab. Invest.* 59, 403, 1988.
6. Meyers, T. W. and Gelfand, D. H., Reverse transcription and DNA amplification by a *Thermus thermophilus* DNA polymerase, *Biochemistry*, 30, 7661, 1991.
7. Nevinney-Stickel, C., Hinzpeter, H., Andrea, A., and Albert, E. D., Non-radioactive typing for HLA-DR1w10 using polymerase chain reaction, dioxigeninlabeled oligonucleotides and chemiluminescene detection, *Eur. J. Immunogenet.*, 18, 323, 1991.
8. Pollard-Knight, D., Read, C. A., Downes, M. J., Howard, L. A., Leadbetter, M. R., Pheby, S. A., McNaughton, E., Syms, A., and Brady, M. A. W. Non-radioactive nucleic acid detection by enhanced chemiluminescence using probes directly labeled with horseradish peroxidase, *Anal. Biochem.*, 37, 84, 1990.
9. Goto, K. and Itoh, T., Detection of *Clostridium piliforme* by enzymatic assay of amplified cDNA segment in microtitration plates, *Lab. Anim. Sci.*, 46(5), 493, 1996.
10. Goto, K. and Itoh, T., Detection of *Bacillus piliformis* by specific amplification of ribosomal sequences, *Jikken Dobutsu*, 43(3), 389, 1994.
11. Dewhirst, F. E., Paster, B. J., Olsen, I., and Fraser, G. L., Phylogeny of the Pasteurellaceae as determined by comparison of 16S ribosomal ribonucleic acid sequences, GenBank Database, National Center for Biotechnology Information, http://www.ncbi.nlm.nih.gov/irx/genbank/-query_form.html, 1997.
12. Kodjo, A., Direct Submission, GenBank Database, National Center for Biotechnology Information, http://www.ncbi.nlm.nih.gov/irx/genbank/query-_form.html, 1997.

chapter seven

Gram-positive cocci

Contents

7.1   Micrococcaceae 114
    7.1.1   *Micrococcus* 114
    7.1.2   *Staphylococcus* 114
        7.1.2.1   Characteristics of infection 114
        7.1.2.2   Characteristics of the agent 116
                7.1.2.2.1   Morphology 116
                7.1.2.2.2   Cultivation 117
                7.1.2.2.3   Isolation sites 118
                7.1.2.2.4   Differentiation and identification 118
7.2   Streptococcaceae 118
    7.2.1   *Streptococcus* 118
        7.2.1.1   Characteristics of the infection 118
                7.2.1.1.1   Epidemiology 118
                7.2.1.1.2   Clinical disease and pathology 120
                7.2.1.1.3   Interference with research 121
        7.2.1.2   Characteristics of the agent 121
                7.2.1.2.1   Morphology 121
                7.2.1.2.2   Media 121
                7.2.1.2.3   Sampling sites 121
                7.2.1.2.4   Immunological identification of isolates .. 121
                7.2.1.2.5   Biochemical identification of isolates 122
    7.2.2   *Enterococcus* 123
    7.2.3   *Aerococcus* 123
    7.2.4   *Gemella* 123
7.3   Peptococcaceae 124
References 124

Important laboratory animal types of Gram-positive cocci belong to either one of the two families Micrococcaceae or Streptococcaceae. Differentiation is done according to Tables 4.1 and 7.1. Micrococcaceae are obligate aerobic

or facultatively anaerobic and grow as large pigmented colonies, while Streptococcaceae are micro-aerophilic and grow as pinpoint colonies. A catalase
test is normally usable for differentiation, as Micrococcaceae are all catalase
positive, while most members of Streptococcaceae are catalase negative.
Some members of the family Streptococcaceae, e.g., aerococci, may show a
weak catalase reaction. The anaerobic cocci, Peptococcaceae, are also found
in laboratory rodents and rabbits, but are of less or unknown importance.
A fourth family, Deinococcaceae, contains one genus, *Deinococcus*, which has
no importance in laboratory animal bacteriology.

7.1   Micrococcaceae

Micrococcaceae in laboratory animals normally belong to one of two genera, *Micrococcus* or *Staphylococcus*. Both grow easily on even simple media
as pigmented colonies, at least 1 mm in size, and are grouped as grapes
(Figure 4.1[1]). Micrococci have an oxidative utilization of carbohydrates
(OF-test) and do not grow under anaerobic conditions, while staphylococci
are fermentative in the OF-test and do grow under anaerobic conditions.
Staphylococci grow in a 12% NaCl broth and are sensitive to Lysostaphin,
which is not the case for micrococci. The main differences are summarized
in Table 7.1.

Apart from these genera two other genera have been described, *Planococcus* and *Stomatococcus*, neither of which are known to have been isolated
from common laboratory animals, although the latter has been isolated
from humans.

7.1.1   Micrococcus

Micrococci are obligate aerobic cocci, which after 24 h of incubation on blood
agar grow as large pigmented colonies. They utilize carbohydrates oxidatively, but in general they utilize far less carbohydrates than staphylococci.
In the microscope they are grouped as pairs or grapes. They can be isolated
from a wide range of animal species worldwide, but they are not known to
have any serious impact on laboratory animals. Different *Micrococcus* species
are identified on the basis of the reactions given in Table 7.2. The use of the
commercial kit, API STAPH (bioMérieux, France), also leads to an identification of micrococci.

7.1.2   Staphylococcus

7.1.2.1   Characteristics of infection

Staphylococci may be found worldwide and in all species of animals. Some
staphylococci are more host specific than others; however, in general, spread
between species, including animal to humans and vice versa, should be
expected. The majority of humans and animals are carriers of staphylococci.
The major pathogen among these bacteria, *Staphylococcus aureus*, is typically

Table 7.1 Differentiation of Different Genera of Gram-Positive Cocci, Which Can Be Isolated from Laboratory Rodents and Rabbits under Aerobic or Microaerophilic Conditions

	Micrococcus	Staphylococcus	Aerococcus	Streptococcus	Enterococcus	Gemella
Aerobic growth	+	+	+	+	+	+
Anaerobic growth	-	+	+	+	+	+
Microaerophilic	-	-	-	-	+	-
OF-test	O	F	F	F	F	F
Microscopic appearance	Pair/tetrades/grapes	Grapes	Pairs/tetrades	Chains	Chains	Single/pairs
Catalase	+	+	Weak or --	-	-	-
Colony pigmentation	+	+	-	-	-	-
Colony size >1 mm	+	+	-	-	-	-
Lysostaphin sensitivity	-	+	-	-	-	-
Growth in						
6.5% NaCl	d	+	+	-	+	-
12% NaCl	-	+	-	-	-	-

Table 7.2 Biochemical Characteristics of Micrococci Found in Rodents and Rabbits

	M. kristinae	M. luteus	M. roseus	M. sedentarius	M. varians
Colony pigmentation	Yellow	Yellow	Red	Cream	Yellow
VP	+	-	-	-	-
Glucose fermentation	+	-	d	+	+
Arginine dihydrolysis	-	-	-	+	-
Nitrate reduction	-	-	+	-	+

found in two thirds of the human population and is also found with a high prevalence in most colonies of laboratory rodents.[1] This is in contrast to wild mice, in which *S. aureus* is rather uncommon. Other types of staphylococci common in laboratory rats and mice include *S. haemolyticus, S. xylosus, S. sciuri*, and *S. cohnii*.[2] The bacteria may be transmitted among hosts in various direct or indirect ways, including passive carriers among animal technicians.

Staphylococcal disease in immune-competent animals is mainly secondary — due to trauma, stress, or the equivalent — and is characterized by pyogenic processes, such as abscesses in bite or surgical wounds, pneumonia in rodents kept in poorly ventilated units, and dermatitis in gerbils kept in too humid bedding. In immune-deficient animals *S. aureus* may be a primary disease-causing agent, e.g., in the nude mouse, in which it causes multiple abscesses. Most pathogenic staphylococci — the most important of which is *S. aureus* — are coagulase-positive, but coagulase-negative staphylococci also may cause disease in laboratory animals; for example, *S. xylosus* is known to cause intestinal disease in mice,[3] dermatitis in gerbils,[4] and pneumonia in immunosuppressed rats.[5] Interference with research is due mainly to the activation of latent infection by stress or immunosuppression; but the presence of abscesses in immune-deficient animals, typically nude mice, also may be hazardous to research.

7.1.2.2   *Characteristics of the agent*

7.1.2.2.1   *Morphology.*   After 24 to 48 h on blood agar staphylococci grow as pigmented colonies at least 1 mm in size. After longer incubation some species may show colonies more than 5 mm in size. At least four different hemolysins exist: α, β, γ, and δ. It should be noted that the type of hemolysis caused by these different hemolysins (Table 7.3) differs from, e.g., streptococcal hemolysin named by equivalent Greek letters. Strains of animal origin produce β-hemolysin more frequently than the other types. In the microscope staphylococci are spherical and are often grouped as grapes (Figure 4.1[1]), although some types, e.g., *S. xylosus*, are not so spherical and do not have a grape-like appearance.

Table 7.3   Characteristics of Staphylococcal Hemolysins

Hemolysin	Characteristics of hemolysis observed on blood agar	Erythrocytes hemolysed		
		Sheep	Rabbit	Horse
α	Wide, sharply demarcated, clear zone	++	+++	−
β	Wide, sharply demarcated, dark zone	+++	+	−
γ	No hemolysis shown on blood agar, should be tested in an erythrocyte broth suspension	+++	+++	−
δ	Narrow, sharply demarcated, clear zone	++	++	+++

7.1.2.2.2 Cultivation. Staphylococci grow easily after direct inoculation on blood agar. If the samples are heavily contaminated a selective agar, such as mannitol salt agar (Table 7.4), may be used for inoculation. If isolation of *S. aureus* is the sole aim Baird Parkers agar (Table 7.5) is of great use.

Table 7.4 Mannitol Salt Agar, a Valuable Medium for the Isolation of Staphylococci[a]

Water	1000 ml
Peptone	10.0 g
Beef extract	1.0 g
NaCl	75.0 g
Mannitol	10.0 g
Phenol red	0.025 g
Agar	15.0 g

Note: pH should be adjusted to 7.2 to 7.4.

[a] *S. aureus* ferments mannitol and produces a golden pigment.

From Røder, B. L., *Substrathåndbogen* [Handbook of Substrates], Seruminstitut, Copenhagen, 1993, 43. With permission.

Table 7.5 A Recipe for Baird Parkers Agar[a]

Water	1000 ml
Peptone 140 (pancreatic digest of caseine)	10.0 g
Peptone 190 (pancreatic digest of gelatin)	5.0 g
Beef extract	5.0 g
Yeast extract	1.0 g
Glycine	12.0 g
Sodium pyruvate	10.0 g
Lithium chloride	5.0 g
Agar	15.0 g

Note: The mixture is boiled to dissolve and autoclaved. After cooling to 50°C 0.1% w/v potassium tellurite and 50 ml egg yolk emulsion are added. pH should be adjusted to 7.0 to 7.2.

[a] The medium is incubated aerobically at 37°C. Lithium chloride prevents growth of Gram-negative bacteria and tellurite prevents growth of coagulase negative staphylococci. *S. aureus* and other coagulase-positive staphylococci form black colonies due to reduction of tellurite, surrounded by a zone, which at first appears clear due to lipo- and proteolysis, but later becomes unclear due to lecithinase and lipase activity. Identification is not 100% specific and should be confirmed by other means. The agar base is commercially available from Gibco (U.S.).

7.1.2.2.3   Isolation sites.   From the diseased animals the organ with the pyogenic process is sampled directly. From healthy animals staphylococci may be isolated from all parts of the respiratory system, the skin, and the genitals.

7.1.2.2.4   Differentiation and identification.   The coagulase test is the first step in the identification process. The vast majority of coagulase-positive staphylococci in laboratory rodents is *S. aureus*. As an alternative or as a supplement to the coagulase test the test for clumping factor, the so-called bound coagulase test, may be used (nearly all coagulase-positive staphylococci show the ability to "clump" erythocytes). The main exception is *S. hyicus*, which is isolated mostly from pigs and never from rodents and rabbits. Commercial kits for the bound coagulase test are available from bioMérieux (France).

If a further identification is needed staphylococci easily may be divided on the basis of simple biochemical characteristics, e.g., fermentation assays, and identified according to Table 7.6. A commercial kit, API STAPH, is available from bioMérieux, France.

Phage typing is a technique especially used for characterization of *S. aureus*. The isolate is incubated with a range of suspensions, each containing one of a number of well-characterized bacteriophages. Each type of *S. aureus* is lysed by certain bacteriophages and not by others.

Molecular probes for identification of *S. aureus* are commercially available from Gen-Probe Inc. (U.S.).

7.2   Streptococcaceae

Streptococcaceae are catalase negative, grow pinpoint colonies, and are grouped in chains (Figure 4.1[2]) or in pairs (Figure 4.1[3]). Streptococcaceae may be divided into the following genera: *Streptococcus, Aerococcus,* and *Gemella*, which are of more or less importance in laboratory animal bacteriology; and *Pediococcus* and *Leuconostoc*, which are unimportant for laboratory animals. The basis for differentiation between genera is given in Table 7.1. All media for cultivation of Streptococcaceae must be enriched, i.e., they should contain serum or the equivalent.

7.2.1   Streptococcus

Streptococci are catalase negative, microaerophilic, grouped in chains (Figure 4.1[2]), and form pinpoint colonies on most agars.

7.2.1.1   Characteristics of the infection

7.2.1.1.1   Epidemiology.   *Streptococcus* spp. of Lancefield's groups A, B, C, and G may be found in laboratory animals all over the world. Group A is uncommon in barrier-housed colonies, while the other types are more

Table 7.6 Biochemical Reactions of Staphylococci Found in Rodents and Rabbits

	S. aureus	S. cohnii	S. haemolyticus	S. hominis	S. lugdunensis	S. saprophyticus	S. sciuri	S. simulans	S. xylosus
Coagulase	+	–	–	–	–	–	–	–	–
Clumping factor	+	–	–	–	–	–	–	–	–
Acid from									
Cellobiose	–	–	–	–	–	–	+	–	–
Lactose	+	–	d	+	–	+	–	+	+
Maltose	+	+	+	+	+	+	+	–	+
Mannitol	+	+	+	–	–	+	+	+	+
Mannose	+	–	–	–	+	–	d	d	+
Sucrose	+	–	+	+	+	+	+	+	+
Xylose	–	–	–	+	–	–	–	–	+
Arginin dihydrolasis	+	–	+	+	–	d	–	+	+/–[a]
Urease	d	d	–	+	d	+	–	+	+/–[a]

[a] Some strains of S. *xylosus*, which are pathogenic in rats,[5] are arginine dihydrolasis positive and urease negative, which are opposite to typical strains of S. *xylosus*.

common, although many colonies may be free of these bacteria, especially hemolytic types. The observed prevalences of a certain species within infected barrier-bred colonies are generally 8 to 10%. *S. pneumoniae* may be found in rats, guinea pigs, rabbits, and, in rare cases, in mice. The agent is more common in guinea pigs than in rats and rabbits. The most common type in guinea pigs is type 19, while types 2, 3, 8, 16, and 19 have been reported in rats. The prevalence within infected colonies may vary from 15 to 55%. Most streptococci, *S. pneumoniae* included, may spread from humans to animals and vice versa. Infection is mostly by droplets through the intranasal route. In barrier-maintained colonies of laboratory animals streptococci are mainly of human origin.

  7.2.1.1.2   Clinical disease and pathology.   Most infections with streptococcal species are nonclinical. It is generally considered of major importance to differentiate between β-hemolytic streptococci, i.e., those which on blood agar form a clear zone around the colony, and the other types, i.e., those which are either nonhemolytic or α-hemolytic, forming a green zone around the colony. This differentiation is based on most pathogenic streptococci being β-hemolytic, although some of the other types also may show some pathogenicity. According to the FELASA guidelines for health monitoring[6,7] (see Chapter 1), infections with streptococci other than *S. pneumoniae* in rodent colonies are only to be reported if these streptococci are β-hemolytic.
  Lancefield's groups A, B, C, and G — *S. zooepidemicus* (group C) is the cause of various pyogenic processes — such as lymphadenitis, submandibular abscess, pneumonia, pleuritis, pericarditis, peritonitis, arthritis, conjunctivitis, opthalmia, mastitis, otitis media, and otitis interna — resulting in high mortality rate in guinea pigs. In rats and mice, species of groups B and G are the most common. Disease is mostly related to group G. The pathological signs resemble those of group C in guinea pigs but are typically much milder with little or no mortality, the most common condition being conjunctivitis. Group A streptococci in rats and mice cause disease equivalent to that caused by group C streptococci in guinea pigs.
  S. pneumoniae — Disease is mostly related to stress due to a poor environment in the animal unit or nutritional deficiencies. In rats the symptoms related to the respiratory system, conjunctiva, and the ears are the most common. The first symptom may be a mucopurulent discharge from the nose, but later the disease progresses and noisy, abdominal respiration occurs. Pathological changes are dominated by fibrin with various grades of focal bronchopneumonia developing into lobar fibrinous pneumonia. In rare cases liver, spleen, kidney, and testes may be involved. In guinea pigs similar symptoms may be present, but often unexpected deaths are the only visible signs of the infection. In rabbits disease often quickly leads to septicemia after signs of dyspnea and depression. Disease due to this agent seldom occurs in mice.

7.2.1.1.3 Interference with research. Animals with subclinical changes in the respiratory system may be more prone to deaths under anesthesia. Stress may induce clinical disease from latent infection with *S. pneumoniae.*

7.2.1.2 Characteristics of the agent

7.2.1.2.1 Morphology. Streptococci grow as small, circular, transparent, so-called pinpoint colonies on blood agar after 24 to 48 h. α-Hemolysis is the most common; β-hemolysis occurs; γ-hemolysis is rare. *S. pneumoniae* appear tiny, smooth, and flat with a small rim of α-hemolysis after 24 h of incubation on blood agar. After 24 h the colonies develop a characteristic ring-like appearance with a raised periphery around a depressed center. In the microscope the cocci are found to be not as round as staphyloccci. Most streptococci occur in chains (Figure 4.1[2]), while *S. pneumoniae,* which are lancet-shaped, are arranged in pairs (Figure 4.1[3]).

7.2.1.2.2 Media. Cultivation may be obtained after direct smear on blood or chocolate agar. Incubation should be 24 to 48 h at 37°C. Most streptococci will grow in an aerobic environment, but cultivation is facilitated by 5% carbon dioxide (microaerophilic cultivation). Columbia blood agar (Difco, U.S. or Gibco, U.S.) is superior to ordinary blood agar. If routine sampling is performed from healthy animals, it should be noticed that streptococci are often overgrown by faster growing organisms. For propagation various broths are available, e.g., soy broth (BBL, U.S.).

7.2.1.2.3 Sampling sites. The sampling site of choice from healthy animals is the nasopharynx sampled through the trachea (see Chapter 2). The nose and the genital organs may be useful as well. For *S. pneumonia* saline washing of the tympanic cavity may be useful, especially in rats.

7.2.1.2.4 Immunological identification of isolates. Streptococci may be grouped within Lancefield's groups. In laboratory animal bacteriology antibodies against types A, B, C, D (enterococci), F, and G are of common use, but several laboratory animal streptococci of less importance are of other types. Commercial kits for Lancefield's groups are available from bioMérieux (France) or Meridian Diagnostics (U.S.), for example. Latex agglutination kits for the identification of *S. pneumoniae* are available from the same suppliers. Identification of the latter in such kits should be regarded as presumptive and should be confirmed by additional means.

Eighty-four different types of *S. pneumoniae* have been identified based on differences in capsular polysaccharide antigens. These may be differentiated by the Neufeld reaction, i.e., visualizing the capsule, which swells when the isolate is mixed with an anticapsular serum.[8]

7.2.1.2.5   Biochemical identification of isolates.   Streptococci may be identified as to their biochemical reactions according to *Bergey's Manual*.[9] Biochemical reactions for relevant laboratory animal species are shown in Table 7.7.

Table 7.7   Identification of Streptococcaceae Found in Rodents and Rabbits after Grouping with Lancefield's Antigens

Group A		
S. pyogenes		
Group B		
S. agalactiae		
Group C	Hemolysis	Trehalose
S. dysgalactiae	α	+
S. equisimilis	β	+
S. zooepidemicus	β	−
Group D	Pyruvate	Arabinose
Enterococcus faecalis	+	−
E. faecium	−	+
Group G		
No further identification		

The most important test for identification of *S. pneumoniae* is the test for sensitivity to optochin, as *S. pneumoniae* produce an autolysin in the presence of surface-active agents, such as bile salts or optochin. The optochin test is also commercially available as discs to be placed upon an agar, from Rosco (Denmark) or Oxoid (U.K.), for example. A 15-mg disc is placed upon a blood agar plate according to the instructions given in Chapter 4. *S. pneumoniae* species will demonstrate a sensitivity zone with a diameter of at least 10 mm. Alternatively, the bile salt sensitivity test may be run. One drop of phenol red is added to 1 ml of a heavy broth culture and NaOH is added until the color turns orange-red. Then two or three drops of sodium taurocholate are added, which after 5 min results in dissolvement of the bacteria — if these are bile salt sensitive.

The API 20 Strep kit from bioMérieux (France) identifies most laboratory animal streptococci in 4 or 24 h.

Molecular probes for identification of group A and B streptococci, as well as *S. pneumoniae* are commercially available from Gen-Probe Inc. (U.S.).

7.2.2   Enterococcus

Enterococci could most easily be described as enteric streptococci, i.e., they are catalase negative, microaerophilic, form pinpoint colonies, and are grouped in chains. Most of them react with Lancefield's antibodies group D and may be divided into two groups: those which are resistant to tellurite and those which are not. *E. faecalis* will grow on tellurite-containing agars, on which it forms large, black colonies. This is routinely used as a selective and indicative method for isolation of enterococci. However, the most common species in small rodents is *E. faecium*, which will not grow in such media. Therefore this system should not be used for isolation of enterococci in laboratory animal bacteriology. The prevalence of enterococci in rodents ranges between 15 and 40%. In rare cases in immune-deficient animals enterococci may play a role in the development of disease, but normally they should be considered as apathogenic or even as probiotic, i.e., they have a positive impact on the intestinal function of the host. This is especially the case in the rat, as enterococci colonize the pars esophagei of the stomach, and from there produce a stream of enterococci that will pass through the intestinal system with the diet and compete with pathogenic bacteria.

Principles for identification are very similar to what has been described for streptococci, and may be performed according to Table 7.7. Alternatively, the commercial kit API 20 Strep (bioMérieux, France) may be applied.

7.2.3   Aerococcus

Aerococci (and pediococci, which are not found in laboratory rodents or rabbits) are grouped in pairs or tetrades and therefore may be taken as micrococci at microscopy. However, as microaerophilic bacteria with little or no catalase reaction, they belong to Streptococcaceae. *A. viridans* is occasionally isolated from the respiratory system of both rodents and rabbits, but any important impact on the animal or its use as a model has not been described. It forms pinpoint colonies on blood agar after 24 h of microaerophilic incubation, with an evident greening in the agar around the colony, especially after longer incubation. Identification may be performed according to Table 7.1 or with the commercial kit API 20 Strep (bioMérieux, France).

7.2.4   Gemella

Gemella in their cultural and morphological characteristics resemble streptococci, i.e., they are catalase negative and form pinpoint colonies. However, although Gram positive, they are easily destained in the Gram staining and may, therefore, falsely be characterized as Gram negative in the initial part of the identification process. They grow easily on blood agar. Two species are of some importance, *G. haemolysans*, which is β-hemolytic, and the non-hemolytic *G. morbillorum*, previously known as *Streptococcus morbillorum*. Both types are occasionally isolated from laboratory rodents, especially rats. They may be difficult to distinguish from *Pediococcus* spp., which, however, are not found in laboratory rodents and rabbits, and therefore are of no importance in this context. Identification may be performed according to Table 7.1 or with the commercial kit API 20 Strep (bioMérieux, France).

7.3 Peptococcaceae

Anaerobic Gram-positive cocci are grouped in one family, Peptococcaceae, which in traditional bacteriology has been divided into five genera: *Coprococcus*, *Peptococcus*, *Peptostreptococcus*, *Ruminococcus*, and *Sarcina*. The commercial kit API 20A (bioMérieux, France) may be useful in the identification process, especially if the isolate happens to be a *Peptostreptococcus* spp. However, it should be noted that these bacteria are poorly characterized in laboratory animals, and therefore identification to a species level is often not possible due to the simple fact that some of these bacteria have not been fully characterized. It is evident that the anaerobic flora has a positive impact on the animal, but it is difficult to make any further statements on positive or negative impact of these bacteria on the use of animals as biomedical models.

References

1. Hansen, A. K., The aerobic bacterial flora of laboratory rats from a Danish breeding centre, *Scand. J. Lab. Anim. Sci.*, 19(2), 37, 1992.
2. Vogelbacher M. and Bohnet, W., Distribution of *Staphylococcus* species in laboratory mice and laboratory rats compared with those found in house mice and Norway rats, in *Harmonization of Laboratory Animal Husbandry*, O'Donoghue, P. N., Ed., Proc. 6th Symp. of the Federation of the European Laboratory Animal Science Associations, June 19–21, Basel, Switzerland, Royal Society of Medicine, London, 1997.
3. Rozengurt, N. and Sanchez, S., Enteropathogenic catalase-negative cocci, in *Welfare and Science*, Bunyan, J., Ed., Proc. 5th Symp. of the European Laboratory Animal Science Associations, June 8–11, Brighton, U.K., Royal Society of Medicine Press, London, 1994, 402.
4. Solomon, H. F., Dixon, D. M., and Pouch, W., A survey of staphylococci isolated from the laboratory gerbil, *Lab. Anim. Sci.*, 40(3), 316, 1990.
5. Detmer, A., Hansen, A. K., Dieperink, H., and Svendsen, P., Xylose-positive Staphylococci as a cause of respiratory disease in immunosuppressed rats, *Scand. J. Lab. Anim. Sci.*, 18(1), 13, 1990.
6. Kraft, V., Deeney, A. A., Blanchet, H. M., Boot, R., Hansen, A. K., Hem, A., van Herck, H., Kunstyr, I., Milite, G., Needham, J. R., Nicklas, W., Perrot, A., Rehbinder, C., Richard, Y., and de Vroy, G., Recommendations for the health monitoring of mouse, rat, hamster, guinea pig and rabbit breeding colonies, *Lab. Anim.*, 28, 1, 1994.

7. Rehbinder, C., Baneux, P., Forbes, D., van Herck, H., Nicklas, W., Rugaya, Z., and Winckler, G., Recommendations for the health monitoring of mouse, rat, hamster, guinea pig and rabbit experimental units, *Lab. Anim.*, 30, 193, 1996.
8. Neufeld, F. and Etinger-Tulcynska, R., Untersuchung über die Pneumokokkenseuche des Meerschweinchens, *Z. Hyg. Infek.*, 114, 324, 1932.
9. Holt, J. G., Krieg, N. R., Sneath, P. H. A., Staley, J. T., and Williams, S. T., *Bergey's Manual® of Determinative Bacteriology*, 9th ed., William & Wilkins, Baltimore, 1994.
10. Koch, F. E., Electivnährboden für Staphylokokken, *Zentralbl. Bakteriol. Parasitenkd. Infektionskr. Hyg. Abt. I Orig.*, 149, 122, 1942.

chapter eight

Gram-positive rods isolated by simple cultivation

Content

8.1   Regular nonacid-fast, nonsporogenic rods 128
      8.1.1   Obligate aerobic regular nonsporogenic rods 128
      8.1.2   Facultatively anaerobic regular nonsporogenic rods 130
              8.1.2.1   *Lactobacillus* 130
              8.1.2.2   *Erysipelothrix* 131
                        8.1.2.2.1   Characteristics of infection 131
                        8.1.2.2.2   Characteristics of the agent 131
              8.1.2.3   *Listeria* 132
                        8.1.2.3.1   Characteristics of infection 132
                        8.1.2.3.2   Characteristics of the agent 133
8.2   Coryneform bacteria 134
      8.2.1   *Corynebacterium* spp. 134
              8.2.1.1   Characteristics of infection 134
              8.2.1.2   Characteristics of the agent 137
                        8.2.1.2.1   Morphology 137
                        8.2.1.2.2   Cultivation 138
                        8.2.1.2.3   Isolation sites 138
                        8.2.1.2.4   Differentiation and identification 138
      8.2.2   *Arcanobacterium* 138
      8.2.3   *Rhodococcus* 139
      8.2.4   *Oerskovia* 139
      8.2.5   *Eubacterium* 139
8.3   Gram-positive rods forming mycelium or hyphae 140
      8.3.1   *Actinomyces* spp. 140
              8.3.1.1   Characteristics of infection 140
              8.3.1.2   Characteristics of the agent 140
8.4   Gram-positive sporogenic rods 141
      8.4.1   *Bacillus* 141

          8.4.1.1   Characteristics of infection..141
          8.4.1.2   Characteristics of the agent..142
    8.4.2 *Clostridium*...143
          8.4.2.1   Characteristics of infection..143
          8.4.2.2   Characteristics of the agent..145
                    8.4.2.2.1   Morphology......................................145
                    8.4.2.2.2   Cultivation.......................................146
                    8.4.2.2.3   Isolation sites.................................146
                    8.4.2.2.4   Differentiation and identification..............146
                    8.4.2.2.5   Toxin detection..............................147
References..150

Gram-positive rods found in rodents and rabbits may basically be divided into those which can be isolated by more or less simple cultivation methods, and those which cannot. This chapter will focus on those bacteria that may be isolated when simple cultivation techniques are performed on samples from rodents and rabbits. Several genera are never or seldom isolated by such simple cultivation, and as such they are of no importance in laboratory animal bacteriology and will not be given any further attention in this book.

The initial step in the identification procedure should be basic characterization according to Chapter 4. It should then be possible to allocate the isolate to one of the groups in Table 8.1. A few additional tests should enable a presumptive diagnosis to a genus level.

8.1   Regular nonacid-fast, nonsporogenic rods

Genera of this group are rod shaped, but exhibit a regular shape with little or no pleomorphism. As this is not a very precise definition care should be taken not to confuse these bacteria with the irregular rods.

All of these bacteria need some kind of complex medium for growth, although it need not be more complex than blood or chocolate agar.

8.1.1   Obligate aerobic regular nonsporogenic rods

From this group only *Kurthia* is isolated from laboratory animals. It is characterized by being obligate aerobic, catalase positive, and by lacking the ability to produce acid from fermentation of carbohydrates, while all other regular nonsporogenic obligate aerobic rods produce lactate from carbohydrates. The cells may appear as chains of rods but may also become coccoid in older cultures. They are readily grown on simple media, e.g., blood agar. The genus *Kurthia* consists of two species: *K. gibsonii* grows at 45°C, and *K. zopfii* does not. The impact of these two organisms on the animals, if any, is not known. Other genera within this group are not known to occur in laboratory rodents and rabbits.

Table 8.1 Grouping of Gram-Positive Rods Which May Be Isolated by Simple Cultivation from Laboratory Rodents and Rabbits

Group in *Bergey's Manual*	Subgrouping			Genera
Regular, nonsporogenic, nonacid-fast	Facultatively anaerobic	Catalase negative	H$_2$S negative	*Lactobacillus*
			H$_2$S positive	*Erysipelothrix*
		Catalase positive		*Listeria*
	Obligate aerobic			*Kurthia*
Coryneform	Catalase positive	Motile		*Oerskovia*
		Nonmotile	Facultatively anaerobic	*Corynebacterium*
			Obligate aerobic	*Rhodococcus*
	Catalase negative	Facultatively anaerobic		*Arcanobacterium*
		Obligate anaerobic		*Eubacterium*
Mycelium or hyphae	Facultatively anaerobic	Catalase positive		*Oerskovia*
		Catalase negative		*Arcanobacterium*
	Obligate anaerobic			*Actinomyces*
	Obligate aerobic	Nonfragmented mycelium		*Streptomyces*
		Fragmented mycelium		*Nocardia*
Sporogenic	Aerobic or facultatively anaerobic	Catalase positive		*Bacillus*
		Catalase negative		*Sporolactobacillus*
	Obligate anaerobic			*Clostridium*

8.1.2   Facultatively anaerobic regular nonsporogenic rods

This group contains three genera of interest in laboratory animal bacteriology: *Erysipelothrix* and *Lactobacillus*, which are catalase negative, and *Listeria*, which is catalase positive. The pathogenic *Erysipelothrix* may differ from the nonpathogenic *Lactobacillus* by microscopy. *Erysipelothrix* are long, slender rods and often filaments are formed, while lactobacilli are straight and occasionally coccoid rods. A simpler differentiation can be based on *Erysipelothrix* producing H_2S, which lactobacilli do not. Test for H_2S production must be performed in Triple Sugar Iron Slant[1] (Table 8.2).

Table 8.2 Triple Sugar Iron Slant for Differentiating
Erysipelothrix rhusiopathiae from Other Gram-Positive
Rods

Meat extract	3 g
Yeast extract	3 g
Peptone	20 g
Glucose	1 g
Lactose	1 g
Sucrose	10 g
$FeSO_4 \cdot 7H_2O$	0.2 g
NaCl	5 g
$Na_2S_2O_3 \cdot 5H_2O$	0.3 g
Agar	20 g
Water	1000 ml
Phenol red, 0.2% in aqueous solution	12 ml

Note: After mixing and sterilization the mixture is dispensed into
tubes to form slopes with deep butts about 3 cm long.

Data from Barrow, G. I. and Fletham, R. K. A., *Cowan and Steel's
Manual for the Identification of Medical Bacteria*, Cambridge University Press, Cambridge, 1993.

8.1.2.1   Lactobacillus

Lactobacilli are Gram-positive rods with a regular shape and often occurring as cocci. They grow on enriched media, e.g., blood or chocolate agar, but often food laboratories will use more sophisticated media including yeast, tomato juice, etc. They form pinpoint colonies and — under aerobic conditions — a greening of the agar will be observed on blood agar. Microaerophilic cultivation is highly recommended. They are normally present in the intestinal tract of laboratory animals and may even be used as probiotics or association flora for rederived animals due to their importance for digestion. However, they are often difficult to isolate from the intestinal tract due to overgrowth from other organisms and are therefore more frequently isolated from the respiratory or genital organs. *L. lactis lactis* is frequently isolated from rodents, but apart from this little is known about specific types and their relation to laboratory animals. The API 50CHL and partly the API 20 Strep and the API 20A kits (bioMérieux, France) will identify lactobacilli to a species level, but identification is difficult for nonspecialized laboratories, and even *Bergey's Manual*[1] does not provide any clear points about species identification.

8.1.2.2   Erysipelothrix

8.1.2.2.1   Characteristics of infection.   In principle, only one species, *E. rhusiopathiae*, should be considered. Another species, *E. tonsillarum*, has been described,[2] but it has not been biochemically differentiated from *E. rhusiopathiae* and it is not known whether it is found in rodents and rabbits, so in this context it does not have any importance. *E. rhusiopathiae* is important in pigs, in which it might be both asymptomatic and cause characteristic symptoms differing between septicemia, endocarditis, arthritis, and/or pathognomonic demarcated skin lesions (erysipelas). From pigs the infection is occasionally transferred to humans, who may develop erysipelas. This problem is not known to occur in humans working with rodents and rabbits. It may be found in the wild rat population and may occasionally be the cause of arthritis in laboratory rats. Rats infected with *E. rhusiopathiae* show an increased activity of β-lymphocytes.[3]

8.1.2.2.2   Characteristics of the agent.   *E. rhusiopathiae* are slender or slightly curved long rods with a tendency to form long filaments. It is easily grown on simple, but enriched media, e.g., chocolate or blood agar. After 24 h of microaerophilic, aerobic, or anaerobic cultivation at 37°C pinpoint colonies are formed. Prolonged incubation does not increase the colony size. Selective enrichment is generally not performed, but Brain Heart Infusion broth with 1% glucose (Gibco, U.S.) is usable for propagation, although other agents may propagate as well from mixed infections. Twenty-two serovars of *E. rhusiopathiae* have been described.[4] The division is made by agar precipitation. Only types 1 and 2 are common.

In healthy animals the nose seems to be the sample site of choice. From affected animals the affected sites should be sampled.

Confirmation of the diagnosis of *E. rhusiopathiae* may be made according to the reactions given in Table 8.3. Alternatively, the API Coryne kit (bioMérieux, France) may be applied, although it should be noted that identification solely based on this kit appears unreliable. Serology is only used by some producers of porcine vaccines in their efforts to describe titers caused by new vaccines. It has not been used in routine screening of farm or laboratory animals. Polymerase chain reaction may be used for direct diagnosis on samples.[5]

Table 8.3 Biochemical Reactions
for *Erysipelothrix rhusiopathiae*

Aerobic growth	+
Anaerobic growth	+
Motility	−
Catalase	−
Gas from glucose	−
Acid from	
Arabinose	−
Galactose	+
Lactose	+
Maltose	+
Mannitol	−
Melezitose	−
Melibiose	−
Raffinose	−
Salicin	−
Sorbitol	−
Trehalose	−
VP	−
NO$_3$ reduction	−
Aesculin hydrolysis	−
Arginin hydrolysis	+

8.1.2.3   Listeria

8.1.2.3.1   Characteristics of infection.   A range of species has been described within this genus, but only one species, *L. monocytogenes*, is known to cause disease in laboratory animals; however, this infection is rather uncommon. It may occur in rare cases in rabbits and guinea pigs, while rats are rather resistant to infection. Although this agent traditionally is known to be pathogenic in mice, it does not play a major role as a spontaneous infection in laboratory mice. The bacterial epidemiological cycle is not fully known. Excretion is thought to be from the mouth, nose, and vagina. It may be found in humans, as well as in various parts of the environment, such as feed and water. According to the FELASA guidelines for health monitoring[6,7] (see Chapter 1), monitoring for *Listeria* infections is not mandatory.

Most infections are subclinical, and disease is related mainly to either contaminated diets or immunodeficiencies of the host. In the latent infected carrier, the agent is found within the macrophages. Septicemia, which in rabbits may be peracute, may lead to death, after which blood and other fluids are accumulated in the abdomen, thorax, and the pericardial sac; hemorrhages and lymph node edemas are also observed during necropsy. In more prolonged cases, various grades of depression and weight loss may be observed and during necropsy gray pinpoint necroses may be found in the liver, spleen, and uterus. Due to the latter, abortions may be caused by infection with

L. monocyctogenes. Meningitis and encephalitis may also be found. Bacterial excretion may be from the digestive, respiratory, or genital systems.

Types of research interference other than those related to disease, death, or pathological changes have not been described.

8.1.2.3.2 *Characteristics of the agent.*

Morphology — *Listeria* is a short, rather thick rod, but especially in young cultures, coccoid forms may be observed. The bacterium is definitely Gram positive, although destaining may occur in older cultures. It is motile at 22°C.

Colony morphology differs slightly between media. *Listeria* colonies are always rather small and often visible only under a stereo microscope. On blood agar both smooth, α-hemolytic and rough, nonhemolytic colonies may be found.

Cultivation — *L. monocytogenes* grows readily on most simple media. All species of *Listeria* will survive and grow when placed at 1° to 5°C in a refrigerator, which will enable the isolation of the agent from contaminated samples due to this kind of selectivity. However, time seldom allows this method to be used in a health monitoring laboratory. Therefore, samples are usually placed in an enriched medium, e.g., Listeria Enrichment Broth commercially available from Merck (Germany). In general, the selective principle is based on the addition of 0.2% thallium acetate or 25 to 75% tellurite salts, and eventually 4 mg/l of nalidixic acid to a serum broth. The broth is incubated at 30°C for 48 h, preferably under microaerophilic conditions. The broth then is streaked on a selective-indicative agar or, alternatively, only an enriched agar. A commercial *Listeria* agar is available from Merck (Germany). Alternatively, PALCAM agar (Table 8.4) may be used.

Table 8.4 Phenol Red Acriflavin Lithium Chloride Columbia Agar with Mannitol (PALCAM Agar) To Be Used for Cultivating *Listeria* spp.[a]

Water	1000 ml	
Peptone	23.0 g	
Starch	1.0 g	
NaCl	5.0 g	
Agar	25.0 g	*Note:* pH is adjusted to
Phenol red	0.08 g	7.2.
Acriflavin	0.005 g	
Lithium chloride	15.0 g	[a] After approximately
Aesculin	0.75 g	48 h *L. monocytogenes*
Ferric ammonium citrate	0.5 g	will form black colonies, which after some
D-Mannitol	10.0 g	time turn green, occa-
Polymyxin B	81.654 i.u.	sionally with a metallic
Ceftazidim	0.02 g	appearance.

Data from Van Netten, P., Van de Moosdijk, A., Curtis, G. D. W., and Mossel, D. A. A., *Int. J. Food. Microbiol.*, 8, 299, 1989.

Isolation sites — From affected animals the affected sites should be sampled. From healthy animals the nose, trachea, cecum, and genitals should be sampled. A suitable procedure is to drop the piece of vagina that is cut off during sampling into an enrichment broth.

Differentiation and identification — Different species of *Listeria* may be differentiated according to fermentation assays (Table 8.5). The kit API Listeria (bioMérieux, France) also identifies *Listeria* species. ELISA has been used to verify the presence of *L. monocytogenes* in food samples, but due to cross-reactions with several other Gram-positive bacteria, this has thus far not been applied in laboratory animal bacteriology. However, an anti-RNA immunoassay, in which as little as 2.5 pg of *Listeria* rRNA can be detected, has been described.[8] Molecular probes for identification of *L. monocytogenes* are commercially available from Gen-Probe Inc. (U.S.).

8.2   Coryneform bacteria

Previously, the genus *Corynebacterium* contained bacteria, which had the one common characteristic that they showed typical V-forms (Figure 4.1[4]), several of which form so-called "Chinese letters." Such bacteria are characterized as coryneforms. However, in other characteristics the members of the traditional *Corynebacterium* genus differed very greatly from one another. More recent systematics have defined corynebacteria as coryneform, facultatively anaerobic, catalase positive, nonmotile rods, while former members not fulfilling these criteria have been placed in other genera. Former members of *Corynebacterium*, which are of importance in laboratory animal bacteriology, are *Arcanobacterium*, which is catalase negative; *Rhodococcus*, which is obligate aerobic; and *Oerskovia*, which may form filaments. Obligate anaerobic species are categorized as *Eubacterium*, which may be found as part of the intestinal flora of rodents and rabbits.

All coryneform species grow easily on simple media, but enriched media such as blood or chocolate agar may be preferred. Coryneform bacteria from rodents and rabbits differ according to Table 8.6. Alternatively, the commercial kit API Coryne (bioMérieux, France) may be used.

8.2.1   Corynebacterium *spp.*

8.2.1.1   *Characteristics of infection*

Corynebacteria are frequently isolated from laboratory rodents, but little is known about precisely which types infect which laboratory animals. Many isolates are typed as corynebacteria but often not identified any further, as rodent corynebacteria are not well described. Some well-characterized *Corynebacterium* spp. described below are known to be isolated from laboratory rodents. As corynebacterial systematics have not been thoroughly studied in laboratory animals, other species, some of which are difficult to identify on a species level, may as well be isolated.

Table 8.5 Differentiation of *Listeria* spp.

| | CAMP test | | Acid from | | | | |
	S. aureus	R. equi	Mannitol	a-methyl-D-mannoside	L-Rhamnose	Sucrose	D-Xylose
L. monocytogenes	+	−	−	+	+	−	−
L. innocua	−	−	−	+	d	d	−
L. seeligeri	+	−	−	d	−	?	−
L. welshimeri	−	−	−	+	d	?	+
L. ivanovii	−	+	−	−	−	d	+
L. grayi	−	−	+	+	−	−	−
L. murrayi	−	−	+	+	d	−	−

Table 8.6 Differentiation of Aerobic and Facultatively Anaerobic Coryneform Bacteria from Laboratory Rodents and Rabbits

| | Reduction of | | β-Galactosidase | Hydrolysis of | | Acid from | | | | |
	Nitrate	Urease		Gelatin	Esculin	Maltose	Sucrose	Mannitol	Xylose	Raffinose
Corynebacterium										
C. kutscheri	+	+	−	−	+	+	+	−	−	−
C. xerosis	+	−	−	−	−	d	+	−	−	−
C. renale	−	+	−	−	−	+	d	−	d	−

continued

Table 8.6 (continued) Differentiation of Aerobic and Facultatively Anaerobic Coryneform Bacteria from Laboratory Rodents and Rabbits

	Reduction of			Hydrolysis of		Acid from				
	Nitrate	Urease	β-Galactosidase	Gelatin	Esculin	Maltose	Sucrose	Mannitol	Xylose	Raffinose
C. pseudodiphtericum	+	+	–	–	–	–	–	–	–	–
C. minutissimum	–	–	–	–	–	+	d	–	–	–
C. urealyticum	–	+	–	–	–	–	–	–	–	–
C. bovis	–	d	+	–	–	–	–	–	–	–
Oerskovia										
O. turbata	d	–	–	+	+	+	+	–	+	–
O. xanthineolytica	d	–	–	+	+	+	+	–	+	+
Arcanobacterium										
A. haemolyticum	–	–	–	–	–	+	d	–	–	?
Rhodococcus										
R. equi	d	+	–	–	–	–	–	–	–	–
Other species	d	d	–	+	–	–	–	–	–	–

C. *kutscheri* has been found worldwide, but during the last decades it has become rather uncommon in laboratory animals, because they are bred and kept in modern facilities. Infection is usually observed only in rats and mice,[9] although the organism has been isolated from guinea pigs and hamsters as well.[10] In immunocompetent rats the agent causes abscess in the superficial tissues and pulmonary emboli, while embolization in the mouse affects joints, liver, and kidney.[11] Latent infections may be activated by certain experimental procedures, which directly or indirectly suppress the immune response of the host.[12] Co-infection with viruses does not seem to activate latent C. *kutscheri* infections,[13] while genetics seems to be involved in resistance and susceptibility of both rats[14] and mice[15] and, therefore, mortality varies between infected colonies. The modes of excretion and spread of the agent are not fully known. Urine and feces from infected animals probably are contaminated. Transplacental infection has been demonstrated experimentally.[16]

C. *renale* may cause urinary calculus in young rats.[17]

Nude mice occasionally suffer from a syndrome characterized by scaling and crusty skin from which coryneform bacteria are isolated. These may be identified as either C. *minutissimum*, C. *bovis*,[18] C. *urealyticum*,[19] or C. *pseudodiphtericum*[20] (syn. *hoffmanii*). The latter has also been associated with conjunctivitis in mice.

No other *Corynebacterium* spp. found in rodents are known to have any impact on the animals.

The number of animals in colonies infected with C. *kutscheri* may be less than 5%, while skin problems in nude mouse colonies might involve more than 80% of the animals. Studies of experimental virus infection may fail due to infection with C. *kutscheri*.[21]

8.2.1.2 Characteristics of the agent

8.2.1.2.1 Morphology. In the microscope all corynebacteria may be found as V-forms (Figure 4.1[4]), which may be arranged as typical "Chinese letters." Although they are all Gram positive, Gram staining may be variable, in particular for C. *kutscheri*.

With C. *kutscheri*, metachromatic granules are observed inside the rods and these granules seem to be Gram positive, even when the culture loses its Gram positivity. After 24 h at 37°C on blood agar, gray-yellow, smooth, nonhemolytic colonies of 0.5 to 1 mm develop, which after longer incubation may grow to 1.5 to 2 mm. On blood tellurite agar the colonies are black.

In C. *renale*, after incubation on blood agar at 37°C for 24 h colonies are round, smooth, white, and opaque. After longer incubation they become dry and granular. The cells are typical Gram-positive coryneforms, often arranged in pallisades, i.e., many cells stacked closely together. Clubforms with metachromatic granules may occur.

C. xerosis and *C. minutissimum* colonies on blood agar are 1 mm, yellow to tan after 24 h at 37°C, but after 48 to 72 h they usually will enlarge to more than 2 mm. In the microscope they are typical corynebacteria.

C. bovis after 48 h on blood agar at 37°C form colonies, which are non-hemolytic, white, smooth, and arching. Coccoid forms may occur.

C. urealyticum colonies are small and gray. Coccoid forms may occur.

8.2.1.2.2 Cultivation. Most rodent species are nonhemolytic, although hemolytic strains of *C. kutscheri* have been described. They do not grow on MacConkey agar. Aerobic incubation at 37°C for 24 to 48 h is suitable.

8.2.1.2.3 Isolation sites. The cecum, genitals, and respiratory system are suitable sites for isolation of Corynebacteria from healthy animals. Clinically affected animals should additionally be sampled from the cervical lymph nodes to search for *C. kutscheri*, or from the renal pelvis to search for *C. renale*, while attempts to isolate *C. minutissimum*, *C. bovis*, and *C. urealyticum* should be performed on samples from the skin of affected animals.

8.2.1.2.4 Differentiation and identification. All corynebacteria are Gram positive, nonmotile, fermentative and catalase positive, grow at 37°C, are negative in Voges-Proskauer test, and do not produce H_2S. In general, corynebacteria from rodents and rabbits are nonhemolytic and produce acid from glucose. Rodent isolates of coryneform bacteria may differ according to Table 8.6. The commercial kit API Coryne (bioMérieux, France) may be applied as well. Infection with *C. kutscheri* in colonies of rats or mice may be diagnosed by demonstrating serum antibodies in ELISA.[22]

8.2.2   Arcanobacterium

Only one species exists, *A. haemolyticum*, formerly designated as *Corynebacterium haemolyticum*. It has occasionally been isolated from the pharynx of various animal species and humans. Its occurrence and impact on rodents and rabbits have not been described in detail.

The morphology of *A. haemolyticum* resembles that of corynebacteria, but it differs from these by being catalase negative. Differentiation from the catalase-negative regular rods may be based on simple morphological characteristics. Microscopy of *Arcanobacterium* shows rods, some of them in V-forms. In older cultures coccoid forms will appear. This is not likely to be confused with *Erysipelothrix*, which are long filamentous rods, but it may be confused with lactobacilli. However, lactobacilli do not occur as V-forms. Furthermore, prolonging the incubation to more than 24 h enlarges the small colonies of *Arcanobacterium*, while the pinpoint colonies of *Lactobacillus* remain small.

8.2.3 Rhodococcus

This genus contains several species, mostly occurring in soil. One species, *R. equi*, formerly designated as *Corynebacterium equi*, is, however, the most common coryneform species in rodents, while it is not clear whether any of the other species have any relation to laboratory animals. *R. equi* is pathogenic in humans and some farm animals, but no impact on infected laboratory animals has been described.

Rhodococcus spp. clearly differ from *Corynebacterium* by being strictly aerobic. Another difference is the extreme slowness or lack of ability of *Rhodococcus* to ferment glucose, which is easily fermented by all corynebacteria from rodents and rabbits. After 24 h of incubation the colonies are rather small, but after longer incubation the colonies turn large, mucoid, and — after some time — red or orange. In young cultures the cells are typical rods with some filaments and several V-forms, but in older cultures cocci may occur.

R. equi may be identified according to Table 8.6. Identification also may be attempted with the API Coryne kit (bioMérieux, France), but this method is not 100% reliable to a species level if *Rhodococcus* spp. other than *R. equi* are suspected. Other *Rhodococcus* species should be considered if the gelatin hydrolysis assay of the kit is positive, although all other characteristics fit with *R. equi*.

8.2.4 Oerskovia

Oerskovia spp. are occasionally isolated from rodents, in which they are not known to have any importance. *Oerskovia* do not clearly differ from *Corynebacterium*. However, *Oerskovia* are motile, and although motile species also occur within the genus *Corynebacterium*, none of those corynebacteria isolated from rodents and rabbits are known to be motile, and therefore differentiation is rather easy in laboratory animal bacteriology. Two species have been defined within this genus, *O. xanthineolytica* and *O. turbata*. Their difference is most easily seen by their ability to grow at 42°C. *O. xanthineolytica* grows at this temperature, while *O. turbata* does not. Additional characteristics are listed in Table 8.6.

8.2.5 Eubacterium

Coryneform bacteria, which are obligate anaerobic, are characterized as *Eubacterium*. Such bacteria may be isolated from the intestines of rodents and, possibly, also from rabbits. At least 34 defined species and 19 unnamed species have been identified, but little is known exactly about which species of *Eubacterium* infects which species of animals. Thus far it is recommended to limit the identification of such bacteria from laboratory animals to a genus level, i.e., "*Eubacterium* spp.," according to Table 8.1. The commercial kit API

20A (bioMérieux, France) may be applied for identification, but only *E. lentum* and *E. limosum* have been considered in the setup of the kit (these species are of some importance in humans, but it is not known whether this is also the case in laboratory animals).

8.3   Gram-positive rods forming mycelium or hyphae

This group contains three genera, *Streptomyces*, *Nocardia*, and *Actinomyces*, the differentiation of which is rather simple according to Table 8.1. *Streptomyces* is not known to infect laboratory animals and will not be given further attention. *N. asteroides* may cause skin and pulmonary lesions in humans and some animals, including rodents and rabbits. Since this has, however, not been reported, this genus is given no further attention. The following text will focus on *Actinomyces*, which is of some importance in rabbits.

8.3.1   Actinomyces *spp.*

8.3.1.1   Characteristics of infection
Actinomyces may cause multiple pyogranulomatous inflammation with nodules in the skin, thorax, abdomen, and even in the central nervous system. Clinical signs may vary according to the affected site. Actinomycosis is rare in rodents today. It may occur in hamsters, in which it typically affects the salivary glands causing liquefactive necrosis, often with sulfur granules. Rabbits are less susceptible, but as laboratory rabbits are often bred and maintained under less strict hygienic regimes than rodents, actinomycosis is most likely to be found in the rabbit, in which it is a chronic disease with clinical symptoms such as diarrhea and weight loss.

8.3.1.2   Characteristics of the agent
Actinomyces is a branching, filamentous, Gram-positive rod. In principle, aerotolerant species occur, e.g., *A. pyogenes* (formerly *Corynebacterium pyogenes*), which causes pyogenic processes in farm animals. However, species isolated from rodents and rabbits show very sparse, if any, growth under aerobic conditions and, therefore, in laboratory animal bacteriology, *Actinomyces* may be regarded as an anaerobic genus, and incubation should be performed as such. Cultivation may be performed on anaerobic chocolate agar (Table 3.2) at 37°C. The inoculated plates are incubated for at least 48 h, inspected under a stereo microscope, reincubated, and then inspected and reincubated every 3 to 5 days for up to 4 weeks.

From affected animals the affected organs should be sampled, while in healthy animals it is often most successful to isolate *Actinomyces* from the nose.

Actinomyces spp. from laboratory animals have not been systematically defined. Species isolated from hamsters show greatest similarity with *A. bovis*, while species isolated from rabbits are more likely to show similarity with *A. israelii*, neither of them being fully identical. They may be differentiated from one another on the basis of the morphology of the colonies. The microcolonies of *A. israelii* are filamentous, while those of *A. bovis* are smooth and nonfilamentous. Fully grown colonies of *A. israelii* are "molar-like," i.e., they are 1 to 2 mm and look like a tooth, while colonies of *A. bovis* have a sharp edge and are smooth. *A. israelii* is a more typical *Actinomyces*, i.e., filamentous and branching, while *A. bovis* may be coryneform. *A. israelii* occasionally grows aerobically, while *A. bovis* never does. Differentiation may be attempted in the kit API 20A (bioMérieux, France).

8.4 Gram-positive sporogenic rods

Sporogenic rods are traditionally divided into *Bacillus*, which are aerobic or facultatively anaerobic, and *Clostridium*, which are strictly anaerobic.

Bacillus spp. are catalase positive. However, catalase-negative, microaerophilic, sporogenic, Gram-positive rods exist and are grouped in the genus *Sporolactobacillus*. Little is known about whether these can be isolated from laboratory animals and they are not given any further attention in this book. However, it should be noted that sporogenic "lactobacilli" may occur. In the same way, a genus *Desulfotomaculum* contains sporogenic, anaerobic, Gram-positive rods which are not sulfate reducing; however, all *Clostridium* spp. are. Because *Desulfotomoaculum* are uncommon in humans and animals this genus is not examined in this book.

One *Clostridium* spp., *C. piliforme*, is one of the most important bacteria in laboratory animal bacteriology, but as yet it cannot be cultivated by traditional techniques; it is described in Chapter 9.

So, in principle, sporogenic bacteria are to be identified as *Bacillus* if catalase positive and able to grow in air, and as *Clostridium* if catalase negative and obligate anaerobic.

8.4.1 Bacillus

8.4.1.1 Characteristics of infection

Only one species, *B. anthracis*, the causative agent of anthrax, is known to be pathogenic. Anthrax is primarily a disease of

- Horses and ruminants, in which it causes acute septicemia
- Humans, in which it causes more localized infection, which may eventually develop into septicemia
- Pigs, in which local, pharyngeal infections may be observed

Mice have previously been used for diagnosis after experimental infection, which led to edema at the injection site, septicemia, and death within 48 h. It is therefore obvious that mice are susceptible to infection with *B. anthracis* and the subsequent development of anthrax. However, it is very unlikely that this should become a major problem in mouse colonies due to worldwide programs to prevent anthrax in other animals and humans.

Bacillus spp. of various kinds are common isolates from laboratory animals, although the animals may not necessarily be infected or associated with these germs in the traditional sense. Spores often enter the animal facilities with the diets, especially if these are not or only partially autoclaved. The spores may then be found in the coat or in the digestive system.

8.4.1.2   Characteristics of the agent

In traditional bacteriological systematics *Bacillus* spp. are characterized as Gram positive. However, in reality some species are more or less Gram negative, while some species are Gram positive in young cultures and Gram negative in older cultures. The ability to form spores is therefore far more important in the identification process than the Gram stainability.

The cells are typical rods, the size of which normally is much larger than rods of other genera. The sides are normally parallel, and some species, e.g., *B. cereus*, form chains. Some species form rather characteristic colonies, which are large (2 to 4 mm), rough, and fatty, but all types of colonies are observed within this genus, and therefore identification of the genus cannot be based on colony morphology.

Bacillus spp. all grow easily on simple, nonenriched media. Psychotrophic species, with temperature optimum between 20° and 25°C, mesophilic with optimum around 30°C, and thermophilic with optimum between 50° and 55°C, exist. Sporulation is best observed after incubation on a sporulation medium, e.g., the one proposed in Table 8.7.

Table 8.7   Medium for Observation of Spores in *Bacillus* spp.[a]

Dehydrated nutrient broth	4 g
Yeast extract	8 g
MnCl, 4H$_2$O	0.01 g
Agar	2 g
Distilled water ad	1000 ml

[a] The medium is incubated at temperatures according to the temperature optimum for the species in question and examined for sporulation twice per day.

The cecum and the trachea are the two most common sites of isolation.

The identification process should always start by characterizing the spores according to Figure 8.1. The easiest way, then, to identify *Bacillus* spp. is through the use of the combination of commercial kits API 50CHL and API 20E (bioMérieux, France), a method primarily based on carbohydrate assimilation. Alternatively, Table 8.8 can be used. *B. anthracis* is closely related

Bacillus   Clostridium

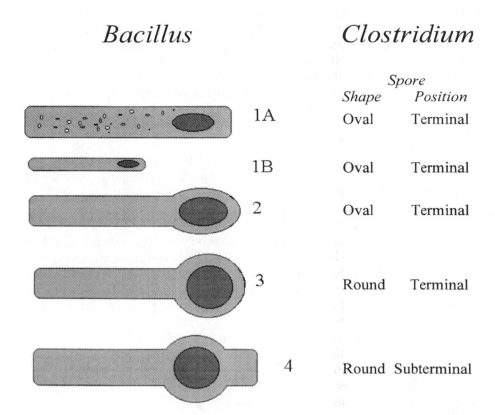

		Spore	
		Shape	Position
	1A	Oval	Terminal
	1B	Oval	Terminal
	2	Oval	Terminal
	3	Round	Terminal
	4	Round	Subterminal

Figure 8.1 Characterization of *Bacillus* spp. according to the spore position, swelling around the spore, and the presence of metachromatic granules in the cytoplasm,[34] and characterization of *Clostridium* spp. according to the shape and position of the spores.

to *B. cereus*. Differentiation between these two is, of course, essential, as the latter is extremely common. First, sensitivity to penicillin is applicable, as *B. anthracis* is sensitive, while all *B. cereus*, except for type 1A, are resistant. *B. anthracis* is nonhemolytic and thereby differs from the hemolytic *B. mycoides*. Finally, *B. cereus* is motile, while *B. anthracis* is not.

8.4.2   Clostridium

8.4.2.1   Characteristics of infection

Pathogenic *Clostridium* spp. are common members of the normal flora. Disease is mainly induced by certain stressors and is mostly related to the production of toxins.

Antibiotic-associated colitis may be induced in hamsters, guinea pigs, and occasionally in rabbits due to treatment with a range of antibiotics, e.g., β-lactams, tetracyclines, clindamycin, gentamicin, and lincomycin.[23] This

Table 8.8 Differentiation of Bacillus spp.

	Spore type	Motility	Anaerobic growth	Glucose gas	Starch hydrolysis	Voges-Proskauer
B. megaterium	1A	d	–	–	+	–
B. thuringiensis	1A	d	–	–	+	–
B. cereus	1A	+	+	–	+	+
B. mycoides	1A	–	+	–	+	+
B. anthracis	1A	–	+	–	+	+
B. licheniformis	1B	+	+	+	+	+
B. subtilis	1B	+	–	–	+	+
B. pumilus	1B	+	–	–	–	+
B. coagulans	2	+	+	–	+	+
B. polymyxa	2	+	+	+	+	+
B. macerans	2	+	+	+	+	–
B. circulans	2	d	d	–	+	–
B. sphaericus	3	+	–	–	–	–

condition is most often caused by *C. difficile*. In guinea pigs it is occasionally associated with *C. perfringens* type E, and in rabbits, especially after treatment with clindamycin, it may be caused by *C. spiroforme*. In all cases the condition is toxin mediated. The toxin A of *C. difficile* plays a more important role in the pathogenesis than toxin B. Both differ from the iota toxins of *C. perfringens* type E and *C. spiroforme*, which are very similar to one another and may be neutralized with the same polyvalent antibodies. The symptoms occur within 10 days after treatment and include ruffled fur, dehydration, diarrhea, and death.

Lethal enteritis in weanling rabbits without a previous history of antibiotic treatment may be caused by iota toxin-producing *C. spiroforme*. Actually, although fairly underestimated, this agent may be extremely important as the cause of rabbit enteritis, probably favored by maldigestion, other infectious agents, and/or nutritional factors. In one investigation the bacterium was detected by Gram stain in 52.4% of 149 cecal samples and iota-like toxin in 7.4% from commercial rabbits showing clinical signs of enteritis complex. From 29 strains of *C. spiroforme* tested, 26 were toxigenic, originating from 24 of 29 rabbitries. In 13.4% of the samples, *C. spiroforme* was present as the only known disease agent. In the other samples, *C. spiroforme* was associated with *Escherichia coli*, *C. piliforme*, rotaviruses, coronavirus, *Eimeria* spp., and cryptosporidia, but there was no significant difference between the presence of these organisms in *C. spiroforme*-positive and -negative samples.[24] Likewise, *C. perfringens* type D may cause lethal enteritis in neonatal mice.

8.4.2.2 Characteristics of the agent

8.4.2.2.1 Morphology. After 48 h of incubation on blood or chocolate agar, colonies of *C. difficile* are 2 to 4 mm in diameter, slightly raised, flat, spreading, and have a rhizoid edge. On CCEY agar (Table 8.9) they are larger. Colony morphology among other clostridial types differ. *C. difficile* mostly, but not always, stain clearly Gram positive. In fact, some other clostridial species, *C. clostridioforme* and *C. sphenoides* mostly, stain Gram negative. Spores of *Clostridium* spp. may differ according to Figure 8.1.

Colonies of *C. perfringens* show a typical double zone of hemolysis on blood agar, the inner zone being fully hemolysed. The colonies are rather large and normally smooth, but may also appear as rough. The cells are large with parallel sides, and spores usually are not observed in standard media.

C. spiroforme may join to form tight coils or spiral configurations (Figure 4.1[11]). This is the main characteristic used for identification of this agent, and *Clostridium* spp. having this configuration may be diagnosed right away as *C. spiroforme*. This is, however, not sufficient for the diagnosis of *C. spiroforme*-mediated diarrhea in rabbits, as only iota toxin-producing *C. spiriforme* may be regarded as pathogenic. Furthermore, it should be noted that *C. spiroforme* does not appear as such in fecal specimens, but rather as semicircular bacteria.[25]

Table 8.9 Cycloserine-Cefoxitin Egg Yolk Agar (CCEY) for
Selective Cultivation of *C. difficile*[a]

Water	1000 ml
Peptone	40.0 g
NaCl	2.2 g
Disodium hydrogen phosphate, 2H$_2$O	6.3 g
Potassium dihydrogen phosphate	1.0 g
Magnesium sulfate, 7H$_2$O	0.2 g
Neutral red	0.03 g
D-Fructose	6.0 g
Cycloserine	0.5 g
Cefoxitin	0.016 g
Pasteurized egg yolk	25 ml
Agar	20.0 g

Note: pH is adjusted to 7.0.

[a] After 48 h of anaerobic incubation colonies of *C. difficile* are 5 to 8
mm, are spreading, and have a rough surface, which under UV
light reflects a yellow-green fluorescence. Furthermore, the colonies smell a little like horse stables, which, however, is not fully
specific for *C. difficile*.

Røder, B. L., *Substrathåndbogen* [Handbook of Substrates], Statens
Seruminstitut, Copenhagen, 1993, 25. With permission.

8.4.2.2.2   Cultivation.   Chocolate agar modified to anaerobic cultivation (Table 3.2) is suitable for initial cultivation, which, of course, must be
performed according to the principles described for anaerobic cultivation. It
is wise to suppress nonsporogenic bacteria in fecal and intestinal samples
by incubating the sample with 50% ethanol for 1 h before inoculation on the
anaerobic chocolate agar. For *C. difficile*, a selective agar such as the CCEY
agar (Table 8.9) may be used.

All clostridia may be grown at 37°C, but *C. perfringens* grows better at 43°C.

8.4.2.2.3   Isolation sites.   In affected animals sampling should be performed from the affected parts of the digestive system. In healthy animals
fecal pellets would be the most appropriate.

A presumptive diagnosis of *C. spiroforme*-mediated diarrhea in rabbits
may be made by directly Gram staining feces from affected rabbits to look
for semicircular bacteria. Feces with such bacteria should be further cultivated according to the principles described above.

8.4.2.2.4   Differentiation and identification.   Some aerotolerant *Clostridium* spp. do exist, the most important one being *C. perfringens*, but as a rule
of thumb, only species that are obligate anaerobic should be subjected to

Table 8.10 Egg Yolk Agar (EYA) for Initial
Examination of *Clostridium* spp.[a]

Water	1000 ml
Peptone	40.0 g
NaCl	2.0 g
Disodium hydrogen phosphate	5.0 g
MgSO$_4$,7H$_2$O	0.02 g
Glucose	2.0 g
Pasteurized egg yolk	25 ml
Agar	25.0 g

Note: pH is adjusted to 7.3.

[a] After 72 to 96 h of incubation the following will be
observable: lecithinase activity; lipase activity; halo
around the colony; oily, shining surface of the colony.

Røder, B. L., *Substrathåndbogen* [Handbook of Sub-
strates], Statens Seruminstitut, Copenhagen, 1993, 33.
With permission.

Clostridium-specific identification in the first place. The spores should be
characterized according to Figure 8.1. A helpful tool is the inoculation of the
isolates on Egg Yolk Agar (Table 8.10).

Identification may be performed according to Table 8.11, which will not
fully identify all species of *Clostridium*. However, it should be possible to
differentiate *C. difficile* and *C. perfringens* clearly from other *Clostridium* spp.
The commercial kit API 20A (bioMérieux, France) will also be helpful in the
identification process. In gas chromatography on norleucine-tyrosine-broth
culture *C. difficile* produces both caproic acid and *p*-cresol, which no other
Clostridium does.[26] *C. perfringens* may, in Table 8.11, be confused with some
closely related species, e.g., *C. clostridioforme*, *C. butyricum*, and *C. septicum*.
C. perfringens is, however, nonmotile, while most other *Clostridium* spp. are
motile. Furthermore, screening for lecithinase activity on Egg Yolk Agar
(Table 8.10) should clearly differentiate *C. perfringens* from other types of
clostridia. Identification by antibodies is possible, by the use of latex agglutination for *C. difficile* or by the Immunocard® (Meridian Diagnostics, U.S.),
i.e., a quick enzymatic antibody test.

8.4.2.2.5 Toxin detection.   Clostridial toxins can be diagnosed directly
in fecal pellets from healthy animals by toxin neutralization assays[27] or
ELISA.[28] *C. difficile* toxin A may also be diagnosed by the Immunocard®
(Meridian Diagnostics, U.S.). PCR has been described as a method for detection of *C. perfringens* toxin genes,[29] which will probably be a method of choice

Table 8.11 Differentiation of *Clostridium* spp.

	Spore type[a]	Egg Yolk Agar		Gelatin hydrolysis	Indole	Acid from					
		Lecith-inase	Lipase			Glucose	Maltose	Lactose	Sucrose	Salicin	Mannitol
C. difficile	OS	−	−	d	−	+	−	−	−	d	+
C. butyricum	OS	−	−	−	−	+	+	+	+	−	−
C. chauvoei	OS	−	−	d	−	+	+	+	+	−	−
C. clostridioforme	OS	−	−	−	d	+	+	d	+	d	−
C. septicum	OS	−	−	+	−	+	+	+	−	+	−
C. histolyticum	OS	−	−	+	−	−	−	−	−	−	−
C. subterminale	OS	−	−	+	−	−	−	−	−	−	−
C. perfringens	OS	+	−	+	−	+	+	+	+	−	−
C. bifermentans	OS	+	−	+	+	+	+	−	−	d	+
C. limosum	OS	+	−	+	−	+	−	−	−	−	+
C. novyi	OS	+	d	+	−	+	d	−	−	−	−
C. sordellii	OS	d	−	+	+	+	+	−	−	−	−
C. sporogenes	OS	−	+	+	−	+	+	−	−	d	−

	Spore[a]									
C. botulinum	O S	–	+	+	–	+	d	–	–	d	–
C. cadaveris	O T	–	–	d	+	+	–	–	–	+	–
C. innocuum	O T	–	–	–	–	+	–	–	d	+	–
C. paraputrificum	O T	–	–	–	–	+	+	+	+	+	+
C. tertium	O T	–	–	–	–	+	+	+	d	+	+
C. ramosum	R/O T	–	–	–	–	+	+	+	+	+	d
C. sphenoides	R S/T	–	–	d	d	+	+	d	d	+	+
C. tetanii	R T	–	–	+	d	–	–	–	–	–	–

[a] O — Oval, R — Round, T — Terminal, S — Subterminal.

for toxins of other *Clostridium* spp. as well, for example the *C. spiroforme* iota toxin, the genetic sequence of which is also known. The iota toxins of *C. perfringens* type E and *C. spiroforme* are so closely related that they will probably be detectable by the same immunological methods, as they may also both be neutralized by *C. perfringens* polyvalent antiserum.

References

1. Weaver, R. E., *Erysipelothrix*, in *Manual of Clinical Microbiology*, Lennette, E.H., Balows, A., Hausier, W.J., Jr., and Shadomy, H.J., Eds., American Society for Microbiology, Washington, D.C., 1985, 209.
2. Takahashi, T., Fujisawa, T., Benno, Y., Tamura, Y., Sawada, T., Suzuki, S., Muramatsu, M., and Mitsuoka, T., *Erysipelothrix tonsillarum* sp.nov. isolated from the tonsils of apparently healthy pigs, *Int. J. Syst. Bacteriol.*, 37, 166, 1987.
3. Ziesenis, A., Röllinger, B., Franz, B., Hart, S., Hadam, M., and Leibold, M., Changes in rat leukocyte populations in peripheral blood, spleen, lymph nodes, and synovia during Erysipelas bacteria-induced polyarthritis, *J. Exp. Anim. Sci.*, 35, 2, 1992.
4. Kucsera, G., Proposals for the designation used for serotypes of *Erysipelothrix rhusiopathiae* (Migula) Buchanan, *Int. J. Syst. Bacteriol.*, 23, 184, 1973.
5. Makino, S., Okada,Y., Maruyama, T., Ishikawa, K., Takahashi, T., Nakamura, M., Ezaki, T., and Morita, H., Direct and rapid detection of *Erysipelothrix rhusiopathiae* DNA in animals by PCR, *J. Clin. Microbiol.*, 32(6), 1526, 1994.
6. Kraft, V., Deeney, A. A., Blanchet, H. M., Boot, R., Hansen, A. K., Hem, A., van Herck, H., Kunstyr, I., Milite, G., Needham, J. R., Nicklas, W., Perrot, A., Rehbinder, C., Richard, Y., and de Vroy, G., Recommendations for the health monitoring of mouse, rat, hamster, guinea pig and rabbit breeding colonies, *Lab. Anim.*, 28, 1, 1994.
7. Rehbinder, C., Baneux, P., Forbes, D., van Herck, H., Nicklas, W., Rugaya, Z., and Winckler, G., Recommendations for the health monitoring of mouse, rat, hamster, guinea pig and rabbit experimental units, *Lab. Anim.*, 30, 193, 1996.
8. Fliss, I., St. Laurent, M., Emond, E., Lemieux, R., Simard, R. E., Ettriki, A., and Pandian, S., Production and characterization of anti-DNA-RNA monoclonal antibodies and their application in Listeria detection, *Appl. Environ. Microbiol.*, 59(8), 2698, 1993.
9. Weisbroth, S. H., *Corynebacterium kutscheri*, in *Manual of Microbiologic Monitoring of Laboratory Animals*, Waggie, K., Kagiyama, N., Allen, A.M., and Nomura, T., Eds., U.S. Department of Health and Human Services, Bethesda, 1994, 129.
10. Amano, H., Akimoto, T., Takahashi, K.W., Nakagawa, M., and Saito, M., Isolation of *Corynebacterium kutscheri* from aged Syrian hamsters (*Mesocricetus auratus*), *Lab. Anim. Sci.*, 41, 265, 1991.
11. Weisbroth, S. H. and Scher, S., *Corynebacterium kutscheri* infection in the mouse. I. Report of an outbreak, bacteriology and pathology of spontaneous infections, *Lab. Anim. Care*, 18, 451, 1968.
12. Weisbroth, S. H., Bacterial and mycotic diseases, in *The Laboratory Rat*, Baker, H. J. and Lindsey, J. R., Eds., Academic Press, New York, 1979, 193.
13. Barthold, S. W. and Brownstein, D. G., The effect of selected viruses on *Corynebacterium kutscheri* in rats, *Lab. Anim. Sci.*, 38, 580, 1988.

14. Suzuki, E., Mochida, K., and Nakagawa, M., Naturally occurring subclinical infection with *Corynebacterium kutscheri* in laboratory rats: strain and age related antibody response, *Lab. Anim. Sci.*, 38, 42, 1988.
15. Pierce-Chase, C. H., Fauve, R. M., and Dubos, R., Corynebacterial pseudo-tuberculosis in mice. I. Comparative susceptibility of mouse strains to experimental infection with *Corynebacterium kutscheri*, *Path. Vet.*, 5, 227, 1964.
16. Juhr, N. C. and Horn, J., Modellinfektion mit *Corynebacterium kutscheri* bei der Maus, *Z.Versuchstierkd.*, 17, 129, 1975.
17. Takahashi, T., Tsuji, M., Kikuchi, N., Ishihara, C., Osanai, T., Kasai, N., Yanaga wa, R., and Hiramune, T., Assignment of the bacterial agent of urinary calculus in young rats by the comparative sequence analysis of the 16S rRNA genes of corynebacteria, *J. Vet. Med. Sci.*, 57(3), 515, 1995.
18. Scanziani, E., Gobbi, A., Crippa, L., Giusti, A.M., Giavazzi, R., Cavalletti, E., and Luini, M., Outbreaks of hyperkeratotic dermatitis of athymic nude mice in northern Italy, *Lab. Anim.*, 31(3), 206, 1997.
19. Richter, C. B., Klingenberger, K. L., Hughes, D., Friedman, H. S., and Schenk man, D. I., D2 coryneforms as a cause of severe hyperkeratotic dermatitis in athymic nude mice, *Lab. Anim. Sci.*, 40(5), 544, 1990.
20. Barthold, S. W., Infectious diseases of mice and rats, http://www.afip.org-/vetpath/POLA/micerat.txt.
21. Barthold, S. W. and Brownstein, D. G., The effect of selected viruses on *Corynebacterium kutscheri*, *Lab. Anim. Sci.*, 38, 50, 1988.
22. Boot, R., Thuis, H., Bakker, R., and Veenema, J. L., Serological studies of *Corynebacterium kutscheri* and coryneform bacteria using an enzyme-linked immunosorbent assay (ELISA), *Lab. Anim.*, 29(3), 294, 1995.
23. Small, J. D., Drugs used in hamsters with a review of antibiotic-associated colitis, in *Laboratory Hamsters*, van Hoosier, G.L., Jr. and McPherson, C.W., Eds., Academic Press, Orlando, 1987, 179.
24. Peeters, J. E., Geeroms, R., Carman, R. J., and Wilkins, T. D., Significance of *Clostridium spiroforme* in the enteritis-complex of commercial rabbits, *Vet. Microbiol.*, 12(1), 25, 1986.
25. Carman, R. J. and Borriello, S. J., Laboratory diagnosis of *Clostridium spiroforme*-mediated diarrhea (iota enterotoxinaemia) of rabbits, *Vet. Rec.*, 113, 184, 1983.
26. Nunez-Montiel, O.L., Thompson, F.S., and Dowell, V.R., Norleucine-tyrosine broth for rapid identification of *Clostridium difficile* by gas-liquid chromatography, *J. Clin. Microbiol.*, 17, 382, 1983.
27. Chang, T. W., Lauermann, M., and Bartlett, J. G., Cytotoxicity assay in antibiotic associated colitis, *J. Infect. Dis.*, 140, 765, 1979.
28. Laughon, B. E., Viscidi, R. P., Gdovin, S. L., Yolken, R. H., and Bartlett, J. G., Enzyme immunoassays for the detection of *Clostridium difficile* toxins A and B in fecal specimens, *J. Infect. Dis.*, 149, 781, 1984.
29. Uzal, F. A., Plumb, J. J., Blackall, L. L., and Kelly, W. R., PCR detection of *Clostridium perfringens* producing different toxins in faeces of goats, *Lett. Appl. Microbiol.*, 25(5), 339, 1997.
30. Barrow, G. I. and Feltham, R. K. A., *Cowan and Steel's Manual for the Identification of Medical Bacteria*, Cambridge University Press, Cambridge, 1993.
31. Van Netten, P., Van de Moosdijk, A., Curtis, G. D. W., and Mossel, D. A. A., Liquid and solid selective differential media for the detection and enumeration of *L. monocytogenes* and other *Listeria* spp., *Int. J. Food Microbiol.*, 8, 229, 1989.

32. George, W. L., Sutter, V. L., and Finegold, S. M., Selective and differential medium for isolation of *Clostridium difficile*, *J. Clin. Microbiol.*, 9, 214, 1979.
33. Røder, B. L., *Substrathåndbogen* [Handbook of Substrates], Statens Seruminstitut, Copenhagen, 1993, 33.
34. Smith, N. R., Gordon, R. E., and Clark, F. E., Aerobic mesophilic sporeforming bacteria, *Agriculture Monograph* 16, U.S. Department of Agriculture, Washington, D.C., 1952.

chapter nine

Gram-positive rods not cultivated by simple techniques

Contents

9.1   *Mycobacterium* spp. 154
      9.1.1   Characteristics of infection...154
      9.1.2   Characteristics of the agent..154
              9.1.2.1   Morphology 154
              9.1.2.2   Cultivation...154
              9.1.2.3   Isolation sites..156
              9.1.2.4   Differentiation and identification.................156
      9.1.3   Safety 156
9.2   *Clostridium piliforme* 158
      9.2.1   Characteristics of infection...158
      9.2.2   Characteristics of the agent..159
              9.2.2.1   Morphology 159
              9.2.2.2   Cultivation 159
              9.2.2.3   Staining...160
              9.2.2.4   Differentiation and identification.................160
              9.2.2.5   Serology..160
              9.2.2.6   Molecular biology...160
References...161

Gram-positive rods not isolated by simple cultivation techniques, and which therefore should not be considered in the identification process following such procedures, are the acid-fast genus *Mycobacterium*, some of which may be isolated by procedures involving specific media and long-term cultivation, and *Clostridium piliforme*, which in contrast to other *Clostridium* spp., grows only in cell cultures, embryonated eggs, or live mice, but, on the other hand, is rather easily diagnosed by serology.

9.1   Mycobacterium *spp.*

9.1.1   Characteristics of infection

The genus *Mycobacterium* includes several species, most of which are saprophytic, while a few are pathogenic for a range of animal species. Typically, the pathogenic mycobacteria cause granulomas with epithelioid cells in various organs, e.g., the lungs, a condition referred to as tuberculosis. In primates and ruminants this is far more common and as such a far more important condition than in small laboratory animals such as rodents and rabbits. These species are to some extent prone to infection and disease, but spontaneous tuberculosis and other types of mycobacterial disease are not likely to become a problem in laboratory rodents and rabbits. *M. avium intracellulare* once has been reported as a spontaneous, latent infection of C57BL/6 mice in a colony, in which 63% of the mice had characteristic mycobacterial lung lesions, i.e., foci with macrophages and multinucleate giant cells.[1] *M. lepraemurium* causes murine leprosy, characterized by granuloma formation around the veins and the capillaries containing so-called "Lepra" cells, i.e. large histiocytic cells with expanded pale cytoplasm and large pale nuclei, which are eccentrically situated. These cells fuse and congregate into nodules.[2] This is, however, a disease of wild rodents and has not been described in laboratory rodents. Infection occurs either aerogeneously or through contaminated diets.

9.1.2   Characteristics of the agent

The slow growth of mycobacteria distinguish them from those Gram-positive rods described in Chapter 8. However, the G + C content of the DNA indicates a relationship with *Corynebacterium, Nocardia,* and *Rhodococcus.*

9.1.2.1   Morphology
Mycobacteria are gracile rods, typically 0.2 to 0.6 µ thick and 1 to 4 µ long. They may branch. The cell morphology varies and is not even typical within one species. Generally, they stain slowly, but when stained, the dyes cannot be evacuated and as such they are Gram positive and acid fast. Colony morphology is not uniform either, but some characteristics are uniform within a species (Table 9.1).

9.1.2.2   Cultivation
Some *Mycobacterium* spp., e.g., *M. lepraemurium* and *M. microti*, cannot be cultivated easily and diagnosis is made on the basis of showing the presence of the agent in characteristic lesions.

Table 9.1 Characteristics of *Mycobacterium* spp. Found in Rodents and Rabbits

	Characteristics of infection	Colony morphology[a]	Sensitivity to infection				
			Rat	Mouse	Hamster	Guinea pig	Rabbit
M. tuberculosis	Generalized	Flat, rough, spreading to irregular periphery	–	–	++	++	+
M. bovis	Generalized	Small, thin, often nonpigmented Raised, rough, later wrinkled and dry	–	–	+	++	++
M. avium/ M. avium intracellulare	Mostly local in lymph nodes and lungs	Thin, transparent, glistening Smooth, entirely rounded Some colonies rough and wrinkled	–	+	+	+	+
M. lepraemurium	Granulomas around veins and capillaries	Not easily cultivated	+	+	–	–	+
M. microti	Local lesions	Not easily cultivated	–	–	–	+	+

a From Cooper and Uyei.[3]

Samples from healthy animals, as well as samples from affected animals in which contamination with other agents is suspected, should be pretreated to destroy the contamination. This may be done by the sulfuric or oxalic acid method[3] (Table 9.2) or methods based on other salts, acids, or bases.[4]

Samples should be inoculated on Löwenstein-Jensen slants[5,6] (Table 9.3). Cultivation for pathogenic *Mycobacterium* spp. should be incubated aerobically at 37°C for at least 7 weeks. Examination for growth must be performed weekly.

A commercial, radiometric detection system, BACTEC AFP, is available from Becton Dickinson (France).

9.1.2.3   Isolation sites

From affected animals the affected sites are sampled. From healthy animals feces should be sampled.

9.1.2.4   Differentiation and identification

Identification on a species level is based on older methods, such as variations in the catalase test, growth on MacConkey agar, and inoculation on laboratory animals, as well as more modern methods such as immunology, high-performance liquid chromatography (HPLC), and polymerase chain reaction (PCR).

Molecular probes for identification of *M. avium* and *M. tuberculosis* are commercially available from Gen-Probe Inc. (U.S.). PCR can be used for a direct diagnosis of infection in humans, and perhaps also in laboratory animals, while serology has not been widely used.

9.1.3   Safety

The U.S. Centers for Disease Control and Prevention (CDC) requires Biosafety Level-2 practices, containment equipment and facilities for preparation of acid-fast smears, and culturing of clinical specimens potentially infected with *M. tuberculosis* or *M. bovis*, provided that aerosol-generating manipulations of such specimens are conducted in a Class I or II biological safety cabinet. The CDC requires Biosafety Level-3 practices, containment equipment and facilities for propagation and manipulation of cultures of *M. tuberculosis* or *M. bovis*, and for animal studies utilizing nonhuman primates experimentally or naturally infected with *M. tuberculosis* or *M. bovis*. Animal studies utilizing guinea pigs or mice can be conducted at Animal Biosafety Level 3. Skin testing with purified protein derivative (PPD) of previously skin-tested-negative laboratory personnel can be used as a surveillance procedure. A licensed attenuated live vaccine (BCG) is available and used in some countries, especially in Western Europe, but not in the U.S.[7]

Table 9.2   Oxalic and Sulfuric Acid Methods for Decontaminating Samples
Previous to Isolation Attempts for *Mycobacterium* spp.

Oxalic acid decontamination	Sulfuric acid decontamination
Materials	
5% oxalic acid	5% sulfuric acid
Sterile PBS	Sterile PBS
4% NaOH	10 ml centrifuge tube
Phenol red indicator	
50 ml centrifuge tube	
Method	
1. Add 5% oxalic acid to the sample (max 10 ml) in the ratio 1:1 in the centrifuge tube.	1. Add 5% sulfuric acid to the sample (3 ml) in the ratio 1:1 in the centrifuge tube.
2. Whirl-mix, and leave at room temperature for 30 min with occasional mixing.	2. Cap the tube and leave it at room temperature for 20 min at constant mixing.
3. Add PBS ad 50 ml.	3. Add PBS ad 10 ml.
4. Cap the tube and mix it several times by inverting.	4. Cap the tube and mix it several times by inverting.
5. Centrifuge 15 min at ≥ 3000 g, decant the supernatant, and add a drop of phenol red to the sediment.	5. Centrifuge 15 minutes at ≥ 3000 g, and decant the supernatant.
6. Neutralize with NaOH.	6. Wash the sediment several times in PBS.
7. Resuspend in PBS and use the suspension for inoculation.	7. Resuspend in PBS and use the suspension for inoculation.

Table 9.3 A Recipe for Löwenstein-Jensen Slants

Sterile, deionized water	612 ml
l-Asparagine	3.6 g
Potassium dihydrogenephosphate	2.4 g
Magnesium citrate	0.6 g
Glycerine	12 ml
Malachite green	0.4 g
Magnesium sulfate, 7H$_2$O	0.24 g
Homogenized egg	1000 ml

Note: pH is stabilized at 6.8 to 7.2. The medium is filled into
tubes as slants.

9.2 Clostridium piliforme

9.2.1 Characteristics of infection

The sole presence of the agent C. *piliforme* in the animal organism, whether causing disease or not, is termed C. *piliforme infection*, while *Tyzzer's disease* describes a condition in which pathological changes are present in the individual due to infection with C. *piliforme*.[8] In mice this highly lethal disease is characterized by multiple focal necroses of the liver, which macroscopically are observed as white spots. Inside the cells of such foci long, thin, slender bacteria are found. These bacteria are also found in huge numbers in the alimentary tract, especially in the ileum and cecum. Tyzzer proposed the name *Bacillus piliformis* for this agent[9] and for many years this name was used, although *Bergey's Manual* never listed this agent as a *Bacillus* spp. Based on RNA analysis the agent has now been allocated to the genus *Clostridium* and has been renamed *Clostridium piliforme*.[10] Tyzzer's disease has also been described in various other species, including rabbits,[11] Mongolian gerbils,[12–14] and Syrian and Chinese hamsters,[15–18] in which the symptoms are similar to mice. In mice, hamsters, and gerbils it should be regarded as a fatal disease characterized by a high mortality in the colony. Also, in rabbits the disease is often peracute with no symptoms prior to finding dead rabbits in the colony. In rats, however, Tyzzer's disease is a mild disease of weanlings connected with megaloileitis, multiple focal necroses of the livers, and single necroses in the myocardium.[19–21] The zoonotic potential of this agent is discussed to some extent, but lately it actually has been isolated from a human immune-deficiency virus-infected patient.[22]

Resistance to development of Tyzzer's disease may be due to genetic traits. Inbred strains of rats carrying MHC-haplotype $RT1.A_1$ seem to be resistant, while a high incidence may be found in rats carrying MHC-haplotype $RT1.A_u$ or $RT1.A_k$.[23,24] Also, ICR mice seem to be more susceptible than other strains of mice.[25]

The organisms probably persist in the intestinal epithelium of healthy animals. The prevalences monitored as the number of individuals with antibodies in infected rat and mouse colonies vary, but it is often more than 50%.[26] In infected rabbit and gerbil colonies the presence of antibodies is usually connected with clinical disease, while antibodies seem to disappear from the colony as mortality declines. In 1990 antibodies against C. *piliforme* were detected in the majority of rat colonies screened, especially in Europe,[27] but today this situation appears very much improved.

A considerable range of different types of research interference may be expected, especially in animals with clinical Tyzzer's disease. This is mostly due to dysfunction of the liver and pathological changes found in the intestines, liver, and myocardium. However, interference may also be observed in nonaffected animals carrying this agent, e.g., the half-life of trimethophrim may be prolonged in mice suffering from acute Tyzzer's disease returning to normal in the recovered mice, while the half-life of warfarin, which is also prolonged during acute Tyzzer's disease, never returns to the values observed in noninfected mice.[28] The agent may cross the placenta barrier.[29,30]

9.2.2 Characteristics of the agent

9.2.2.1 Morphology

C. *piliforme* is a thin rod approximately 0.3 to 0.5 m wide and 8 to 10 m long. In spore-forming strains the spores may be seen as thickenings in the end, which give the agent the shape of a drumstick. The spores are elliptical and are often referred to as "rocket spores." They may be stained by spore staining (Table 4.6), but the simplest way to find them is simply to look for them in an immunofluorescence-stained smear, in which they are easily observed. In histological slides bacterial cells are grouped as bundles (Figure 4.1.[5]).

9.2.2.2 Cultivation

Attempts to cultivate the organism by ordinary bacteriological methods have thus far been unsuccessful.[31] Cell cultures, embryonated eggs, or live mice are necessary for propagation. For ethical reasons, priority should be given to propagation in cell cultures.

Cell lines usable for the propagation of C. *piliforme* include primarily the mouse-embryo fibroblast cell line, 3T3, while a more limited growth is obtained in cell lines of mouse–connective-tissue origin (L-929) and mouse liver origin (NCTC 1469). An initial decrease in the number of bacteria is noted after inoculation of the cell layer with $2.6 \cdot 10^5$ organisms followed by a peak bacterial count at 48 h.[32] Embryonic chimp liver cell cultures have also been used by some.

As an alternative the suspension may be inoculated aseptically into embryonated eggs. Eggs 6 to 9 days old are inoculated in the yolk sac with infected liver suspension prepared as described below. The viability is determined by candling three times per day. Embryonic death within 24 h is disregarded. Yolk sac from embryos dying later can be harvested and frozen at –80°C. A 20% PBS suspension should be examined for the presence of the agent.[33]

For *in vivo* propagation mice are given 3 mg prednisolone, 6 mg sulfadoxine, and 1 mg trimethoprim subcutaneously once per day for two consecutive days. On day 3 a sterile suspension of infected liver tissue in PBS (1:10) is given intravenously to the mice. The medication is repeated. Then the mice must be observed carefully. Affected mice must be euthanized immediately and the liver sampled under aseptic conditions. Only livers with white spots should be used. A piece is sampled and stained by immunofluorescence as described in Chapter 5. If C. *piliforme* is not found on the slide a new liver must be sampled. Livers with the agent may be used for preparing further slides for serology or may be frozen for later use.

9.2.2.3 Staining

Smears of ileum, liver, or myocardium may be prepared and stained either by immunofluorescence or PAP techniques as described in Chapter 5. Samples from the ileum are made by cutting a piece of approximately 1 cm of the ileum, inverting it, and washing the inner surface with distilled water. Then smears are made as described in Chapter 5. A simple staining may be prepared from Giemsa solution. The slide is fixed for 20 s in absolute methanol and then stained for 2 min in undiluted Giemsa solution (Merck, Germany). Using a high titer antibody immunofluorescence staining is superior to both PAP and Giemsa staining. Gram staining is only of limited use for staining this agent.

9.2.2.4 Differentiation and identification

Experimental infection, protein analysis, Western immunoblotting technique, immunodiffusion, and other immunological techniques may be applied to show differences between strains from various animals. There are only slight differences between rat and mouse strains, while major differences between rat strains and strains of other animal species have been observed.[34–37] Competitive ELISA may be used to detect isolate-specific antibodies in sera from infected animals.[38]

9.2.2.5 Serology

Both immunofluorescence assay as well as ELISA are usable for screening for antibodies. The antigen is commercially available from Harlan (U.K.). Care should be taken not to use too low a cutoff value, and testing should be performed primarily on animals more than 10 weeks old. Young germ-free animals may, if infected, respond otherwise to the infection than by producing a high titer of antibodies.[39] High titers may develop as early as 10 days postinfection and reach a maximum within 25 days.[40]

9.2.2.6 Molecular biology

A simple enzymatic assay system using reverse transcription polymerase chain reaction (RT-PCR) products in a microtitration plate has been developed to detect *C. piliforme* in tissue specimens. The RT-PCR is performed with a biotin-labeled primer directed in a selected area of the *C. piliforme* 16S rRNA gene. The amplified cDNA is hybridized to an alkaline phosphatase-labeled DNA probe in a microtube, and the mixture is applied to a microtitration well precoated with streptavidin. Hybridization signals in the microtitration plate are visualized by reaction with substrate for the alkaline phosphatase. The system may detect the RNA from as few as ten organisms. The system may also detect organisms from the liver, mesenteric lymph nodes, and ileum of experimentally infected mice. In the ileum the *C. piliforme* gene may be detected from 1 to 21 days after infection.[41]

References

1. Waggie, K. S., Wagner, J. E., and Lentsch, R. H., A naturally occurring outbreak of *Mycobacterium avium-intracellulare* infections in C57BL/6N mice, *Lab. Anim. Sci.*, 33, 249, 1983.
2. Benischke, K., Garner, F. M., and Jones, T. C., *Pathology of Laboratory Animals*, Springer-Verlag, New York, 1978.
3. Corper, H.J. and Uyei, N., Oxalic acid as a reagent for isolating tubercle bacilli and a study of the growth of acid-fast non-pathogens on different mediums with their reactions to chemical reagents, *J. Lab. Clin. Med.*, 15, 348, 1930.
4. Nolte, F. S. and Methchock, B., Mycobacterium, in *Manual of Clinical Microbiology*, Murray, P. R., Baron, E. J., Pfaller, M. A., Tenover, F. C., and Yolken, R. H., Eds., ASM Press, Washington, D.C., 1995, chap. 31.
5. Löwenstein, E., Die Züchtung der Tuberkelbazillien aus dem strömende Blute, *Zentralb. Bakteriol. Parasitenkd. Infektionskr. Hyg. Abt. I. Orig.*, 120, 127, 1931.
6. Jensen, K. A., Reinzüchtung und Typenbestimmung von Tuberkelbazillen-stämmen. Eine Vereinfachung der Methoden für die Praxis, *Zentralb. Bakteriol. Parasitenkd. Infektionskr. Hyg. Abt. I. Orig.*, 125, 222, 1932.
7. Office of Health and Safety, Centers for Disease Control and Prevention, Laboratory Biosafety Level Criteria, http://www.cdc.gov/od/ohs/biosfty/bmbl/section3.htm.
8. Committee on Infectious Diseases of Mice and Rats, *Infectious Diseases of Mice and Rats*, National Academy Press, Washington, D.C., 1991, 127.
9. Tyzzer, E. E., A fatal disease of the Japanese Waltzing mouse caused by a spore-bearing bacillus (*Bacillus piliformis* N.SP.), *J. Med. Res.*, 37, 307, 1917.
10. Duncan, A. J., Carman, R. J., Olsen, G. J., and Wilson, K. H., Assignment of the agent of Tyzzer's disease to *Clostridium piliforme* comb. nov. on the basis of 16S rRNA sequence analysis, *Int. J. Syst. Bact.*, 43(2), 314, 1993.
11. Allen, A. M., Ganaway, J. R., Moore, T. D., and Kinard, R. F., Tyzzer's disease syndrome in laboratory rabbits, *Amer. J. Path.* 46, 859, 1965.
12. Carter, G. R., Whitenack, D. L., and Julius, L. A., Natural Tyzzer's disease in Mongolian gerbils (*Meriones unguiculatus*), *Lab. Anim. Care*, 19, 648, 1969.
13. White, D. J., and Waldron, M. M., Naturally-occurring Tyzzer's disease in the gerbil, *Vet. Rec.*, 85, 111, 1969.
14. Port, C. D., Richter, W. D., and Moise, S. M., Tyzzer's disease in the gerbil, *Lab. Anim. Care*, 20, 109, 1970.
15. Takasaki, Y., Oghiso, Y., Sato, K., and Fujiwara, K., Tyzzer's disease in hamsters, *Jpn. J. Exp. Med.*, 44, 267, 1974.
16. Nakayama, M., Saegusa, J., Itoh, K., Kiuchi, Y., Tamura, T., Ueda, K., and Fujiwara, K., Transmissible enterocolitis in hamsters caused by Tyzzer's organism, *Jpn. J. Exp. Med.*, 45, 33, 1975.
17. Zook, B. C., Albert, E. N., and Rhorer, R. G., Tyzzer's disease in the Chinese hamster (*Cricetulus griseus*), *Lab. Anim. Sci.* 27, 1033, 1977.
18. Zook, B. C., Huang, K., and Rhorer, R. G., Tyzzer's disease in Syrian hamsters, *J. Amer. Vet. Med. Assoc.*, 171, 833, 1977.
19. Geil, R. G., Davis, D. L., and Thompson, S. W., Spontaneous ileitis in rats: a report of 64 cases, *Amer. J. Vet. Res.*, 22, 932, 1961.

20. Hottendorf, G. H., Hirth, R. S., and Peer, R. L., Megaloileitis in rats, *J. Amer. Vet. Med. Assoc.*, 155, 1131, 1969.

21. Jonas, A. M., Percy, D. H., and Craft, J. C., Tyzzer's disease in the rat. Its possible relationship with megaloileitis, *Arch. Path.*, 90, 516, 1970.

22. Smith, K. J., Skelton, H. G., Hilyard, E. J., Hadfield, T., Moeller, R. S., Tuur, S., Decker, C., Wagner, K. F., and Angritt, P., *Bacillus piliformis* infection (Tyzzer's disease) in a patient infected with HIV-1: confirmation with 16S ribosomal RNA sequence analysis, *J. Am. Acad. Dermatol.*, 34, 343, 1996.

23. Hansen, A. K., Svendsen, O., and Møllegaard-Hansen, K. E., Epidemiological studies of *Bacillus piliformis* infection and Tyzzer's disease in laboratory rats, *Z. Versuchstierkd.*, 33, 163, 1990.

24. Hansen, A. K., Dagnœs-Hansen, F., and Møllegaard-Hansen, K. E., Correlation between megaloileitis and antibodies to *Bacillus piliformis* in laboratory rat colonies, *Lab. Anim. Sci.*, 42(5), 449, 1992.

25. Fujiwara, K., Tyzzer's disease, *Jpn. J. Exp. Med.*, 48, 467, 1978.

26. Hansen, A. K., Skovgaard-Jensen, H. J., Thomsen, P., Svendsen, O., Dagnœs-Hansen, F., and Møllegaard-Hansen, K. E., Rederivation of rat colonies seropositive to *Bacillus piliformis* and the subsequent screening for antibodies, *Lab. Anim. Sci.*, 42(5), 444, 1992.

27. Kraft, V. and Meier, B., Seromonitoring in small laboratory animal colonies. A five year survey: 1984–1988, *Z. Versuchstierkd.*, 33, 29, 1990.

28. Friis, A. S. and Ladefoged, O., The influence of *Bacillus piliformis* (Tyzzer) infections on the reliability of pharmacokinetic experiments in mice, *Lab. Anim.*, 13, 257, 1979.

29. Friis, A. S., Demonstration of antibodies to *Bacillus piliformis* in SPF colonies and experimental transplacental infection by *Bacillus piliformis* in mice, *Lab. Anim.*, 12, 23, 1978.

30. Friis, A. S., Studies on Tyzzer's disease: transplacental transmission by *Bacillus piliformis* in rats, *Lab. Anim.*, 12, 23, 1978.

31. Thunert, A., Is it possible to cultivate the agent of Tyzzer's disease *(Bacillus piliformis)* in cellfree media?, *Z. Versuchstierkd.*, 26(4), 145, 1984.

32. Spencer, T. H., Ganaway, J. R., and Waggie, K. S., Cultivation of *Bacillus piliformis* (Tyzzer) in mouse fibroblasts (3T3 cells), *Vet. Microbiol.*, 22(2–3), 291, 1990.

33. Fries, A. S., Studies on Tyzzer's disease: isolation and propagation of *Bacillus piliformis*, *Lab Anim.*, 11, 75, 1977.

34. Fujiwara, K., Kurashina, H., Magaribuchi, T., Takenaka, S., and Yokoiyama, S., Further observations on the difference between Tyzzer's organisms from mice and those from rats, *Jpn. J. Exp. Med.*, 41(2), 125, 1971.

35. Fujiwara, K., Kurashina, H., Magaribuchi, T., Takena, S., and Yokoiyama, S., Further observations between Tyzzer's organisms from mice and those from rats, *Jpn. J. Exp. Med.*, 43, 307, 1973.

36. Fujiwara, K., Nakayama, M., Nakayama, H., Toriumi, W., Oguihara, S., and Thunert, A., Antigenic relatedness of *Bacillus piliformis* from Tyzzer's disease occurring in Japan and other regions, *Jpn. J. Vet. Sci.*, 47(1), 9, 1985.

37. Riley, L. K., Besch-Williford, C., and Waggie, K. S., Protein and antigenic heterogeneity among isolates of *Bacillus piliformis*, *Infect. Immun.*, 58(4), 1010, 1990.

38. Boivin, G. P., Hook, R. R., Jr., and Riley, L. K., Development of a monoclonal antibody-based competitive inhibition enzyme-linked immunosorbent assay for detection of *Bacillus piliformis* isolate-specific antibodies in laboratory animals, *Lab. Anim. Sci.*, 44(2), 153, 1994.
39. Hansen, A. K., Andersen, H. V., and Svendsen, O., Studies on the diagnosis of Tyzzer's disease in laboratory rat colonies with antibodies against *Bacillus piliformis (Clostridium piliforme)*, *Lab. Anim. Sci.*, 44(5), 424, 1994.
40. Fries, A. S., Studies on Tyzzer's disease: a long term study of the humoral antibody response in mice, rats and rabbits, *Lab. Anim.*, 13, 37, 1979.
41. Goto, K. and Itoh, T., Detection of *Clostridium piliforme* by enzymatic assay of amplified cDNA segment in microtitration plates, *Lab. Anim. Sci.*, 46(5), 493.

chapter ten

Facultatively anaerobic Gram-negative bacteria

Contents

10.1 Enterobacteriaceae 166
    10.1.1 Characteristics of infection..166
        10.1.1.1 *Escherichia*..167
        10.1.1.2 *Citrobacter* 167
        10.1.1.3 *Salmonella* 168
        10.1.1.4 *Klebsiella*...168
        10.1.1.5 *Enterobacter*...168
        10.1.1.6 *Proteus* 168
        10.1.1.7 *Morganella*...168
        10.1.1.8 *Yersinia* 169
        10.1.1.9 Other species 169
    10.1.2 Characteristics of the agent...169
        10.1.2.1. Sampling and cultivation 169
        10.1.2.2 Identification...171
    10.1.3 Safety 176
10.2 Pasteurellaceae 177
    10.2.1 *Pasteurella* 177
        10.2.1.1 Characteristics of infection...................................177
        10.2.1.2 Characteristics of the agent..................................177
            10.2.1.2.1 Morphology 177
            10.2.1.2.2 Cultivation 178
            10.2.1.2.3 Isolation sites....................................178
            10.2.1.2.4 Differentiation and identification..............178
            10.2.1.2.5 Serology...178
    10.2.2 *Actinobacillus* 179
        10.2.2.1 Characteristics of infection...................................179
        10.2.2.2 Characteristics of the agent..................................180
            10.2.2.2.1 Morphology 180

                    10.2.2.2.2 Cultivation 180
                    10.2.2.2.3 Isolation sites...180
                    10.2.2.2.4 Differentiation and identification..............180
                    10.2.2.2.5 Serology..181
                    10.2.2.2.6 Molecular biology......................................181
          10.2.3 *Haemophilus* 181
                    10.2.3.1 Characteristics of infection.............................181
                    10.2.3.2 Characteristics of the agent...........................181
     10.3 *Streptobacillus moniliformis* 182
          10.3.1 Characteristics of infection...182
          10.3.2 Characteristics of the agent...184
                    10.3.2.1 Morphology 184
                    10.3.2.2 Cultivation...184
                    10.3.2.3 Isolation sites..184
                    10.3.2.4 Differentiation and identification................184
                    10.3.2.5 Serology...184
     10.4 Vibrionaceae..185
     References..186

Facultatively anaerobic Gram-negative bacteria isolated from rodents and rabbits in routine bacteriology usually belong to the families Enterobacteriaceae, Pasteurellaceae, or Vibrionaceae. In the latter family only *Aeromonas* is a common finding. The three families differ as indicated in Table 4.2. As the table shows, it is rather simple to differentiate the three by microscopic morphology, oxidase, and motility. *Streptobacillus moniliformis*, the cause of rat bite fever, also falls within this category, but it can easily be differentiated from other facultatively anaerobic Gram-negative bacteria, as it is morphologically different from the other members, not easily isolated on ordinary media after 24 h of incubation, and catalase negative.

10.1 Enterobacteriaceae

Enterobacteriaceae are catalase positive, oxidase negative, and the carbohydrate utilization is fermentative. Both motile and nonmotile species occur.

10.1.1 Characteristics of infection

Some Enterobacteriaceae, e.g., *Proteus, Morganella,* and to some extent *Escherichia,* are commensal members of the intestinal flora, while others, e.g., *Salmonella* and *Yersinia,* are highly undesirable facultative pathogens in laboratory animal colonies.

10.1.1.1 Escherichia

Most rodents, with the important exceptions of hamsters and guinea pigs, carry *Escherichia coli* as a normal inhabitant in the cecum and the large intestines. In rabbits, in which it is not a normal inhabitant but an important pathogen, and also in species in which it is a normal inhabitant, it may cause disease in stressed animals. Enteritis is the most common affection caused by *E. coli*, but it may be involved in other types of disease, e.g., cystitis, pyelonephritis, metritis, pneumonia, and local inflammation.

In rabbit colonies outbreaks of colibacillosis may be connected with high losses during a period, the main symptom being diarrhea. After a period most outbreaks seem to stop by themselves. Often *E. coli* is only secondary and other infections, such as rotavirus, are the main pathogen. The pathogenic effect is mainly caused by toxin production, but in contrast to humans and pigs little is known about which serotypes are the most common pathogens in rabbits, and therefore vaccination with commercially available pig vaccines is of little use in a rabbit colony.

In guinea pigs fatal enteropathies may be caused by *E. coli* after fecal-oral transmission in poorly housed young animals. In lactating females mastitis is occasionally caused by *E. coli*. In rat colonies *E. coli* can be isolated from approximately 70% of cecal samples, while intestinal samples from moribund rabbits lead to isolation in nearly 100% of the cases if *E. coli* is the cause. Other *Escherichia* spp. are not common inhabitants of rodents and rabbits.

10.1.1.2 Citrobacter

Citrobacter spp. can be isolated from the intestines and genitals of a range of animal species. Main species isolated from rodents are *C. freundii* and *C. diversus*. In particular, *C. freundii* is a rather common finding in hamsters and guinea pigs, while *C. diversus* is most common in rats. In most cases these infections do not cause any symptoms and the observed prevalences are rather low. In extreme cases *C. freundii* may cause epizootic pneumonia and enteritis with high mortality in guinea pig colonies.[1] A third species, *C. rodentium*, formerly known as either *C. freundii* type 4280[2] (due to its identification number in the commercial kit API 10E), *C. freundii* type ANL,[3] or *C. freundii* type Ediger,[4] is of major importance as an intestinal pathogen in male mice. Symptoms are most common in sucklings and weanlings in which rectal prolapses, diarrhea, and dehydration are observed. Feeding and genetic factors influence both morbidity and mortality, but generally high losses are not common. Prevalence rates are normally rather low, i.e., often the agent is detectable in less than 5% of the individuals from an infected colony. The condition resembles a common syndrome in certain strains of transgenic mice, but in these mice *C. rodentium* is not necessarily involved. *C. rodentium* may raise the susceptibility of the animal to some types of chemical carcinogenesis.[5]

10.1.1.3 Salmonella

Salmonella spp. infects all species of warm-blooded animals. Until the introduction of barrier-protected breeding systems in the 1960s, *Salmonella* commonly ruined research projects involving the use of rodents, especially if the project involved the use of mice. In mice and rats, in which *S. typhimurium* is the most common, clinical disease is characterized by various grades of diarrhea. During the first period of an outbreak a number of sudden deaths due to septicemia occur, but afterward the conditions turn chronic, and a number of animals will act as persistent carriers. In the highly susceptible guinea pig, *S. enteritidis* is the most common species of *Salmonella*. Symptoms are similar to those in rats and mice, and, furthermore, in breeding colonies abortions may occur. Rabbits may suffer from salmonellosis, but this is rather uncommon. In colonies infected with *S. typhi murium* or *S. enteritidis* the prevalences observed mostly range above 50%, but several other species of *Salmonella* may be introduced through contaminated diets, staff, etc., and in such cases the prevalences are often extremely low, i.e., *Salmonella* may be isolated from a single individual, but the diagnosis is impossible to confirm by additional isolations. Uterine infections — probably without passage of the placenta barrier — have been described for *Salmonella* spp.[6]

10.1.1.4 Klebsiella

Four species — *K. orthinolytica*, *K. oxytoca*, *K. planticola*, and *K. pneumoniae* — may be isolated from both rodents and rabbits. Generally, prevalences are rather low, i.e., 20% or less, and clinical affection is not observed. In mice and guinea pigs, pneumonia and deaths due to septicemia are observed in rare cases. High prevalences may be due to uncritical antibiotic treatments in the colony.[7]

10.1.1.5 Enterobacter

Five species — *E. aerogenes*, *E. agglomerans*, *E. cloacae*, *E. gergoviae*, and *E. sakazakii* — are known to be isolated from laboratory rodents. The most common is *E. cloacae*, which may show prevalences of 25 to 30%. They normally do not cause disease, except for *E. cloacae*, which is a common cause of pneumonia in newborn rabbits.

10.1.1.6 Proteus

Only one species, *P. mirabilis*, is isolated from rodents. Prevalences vary from very low to close to 100%. High prevalences may be caused by uncritical antibiotic treatments in the colony.[4] *Proteus* infection is not associated with disease.

10.1.1.7 Morganella

One species, *M. morganii*, is occasionally isolated from a low number of individuals in rodent colonies. No disease symptoms are known to occur due to this kind of infection.

10.1.1.8 Yersinia

Only one species, *Y. pseudotuberculosis*, is of interest in rodents and rabbits. It may be isolated from all species of rodents and rabbits, although this seldom occurs. It causes so-called pseudotuberculosis in guinea pigs and in rare cases in rabbits. It should be noted that pseudotuberculosis in rats and mice is not caused by *Yersinia* but by *Corynebacterium kutscheri* (see Chapter 8).

Classical pseudotuberculosis in guinea pigs and rabbits is characterized by wasting disease and diarrhea progressing until death after a course of 4 weeks. In both species acute septicemias may, under certain circumstances, lead to death within 48 h.

10.1.1.9 Other species

Other species — *Edwardsiella, Hafnia, Klyuvera, Providencia,* and *Serratia* — may be isolated from rodents but are not known to be correlated with any clinical conditions.

10.1.2 Characteristics of the agent

10.1.2.1 Sampling and cultivation

Morphological descriptions of important members of Enterobacteriaceae are given in Table 10.1. Enterobacteriaceae are easily cultivated on either

Table 10.1 Colony Morphology of Important Enterobacteriaceae in Rodents and Rabbits after Incubation on Blood Agar

Species	Colony morphology (24 h)	Bacterial morphology
Citrobacter spp.	2- to 3-mm large, round colonies with a smooth edge and surface	Short (0.6–1.5 µm), slightly pleomorphic rods
Enterobacter cloacae	Large 2- to 4-mm mucoid colonies (less mucoid than *Klebsiella*)	Long, straight rods, 0.6–1.0 µm wide and 1.2–3.0 µm long
Escherichia coli	2- to 3-mm large, round colonies with a smooth edge and surface	Coliform rods (Figure 4.1[7]), but cocoid forms may occur
Proteus mirabilis	Swarming in waves to cover the whole agar	Coliform rods (Figure 4.1[7])
Klebsiella pneumoniae	Large 3- to 4-mm mucoid, irregular colonies	Plump rods with a thick capsule, some individual cocci may be observed
Morganella morganii	2- to 3-mm flat colonies	Coliform rods
Salmonella spp.	2- to 3-mm flat, circular, gray colonies	Long (2–3 µm) rods
Yersinia pseudotuberculosis	0.5- to 2-mm smooth colonies with no clear edges (30°C)	Pleomorphic rods with rounded ends, staining is irregular

blood agar or MacConkey agar[8] (Table 10.2). If blood agar is used a detergent for prevention of *Proteus* swarming should be included in the recipe (see Chapter 3). Enterobactericeae from laboratory animals easily grow after 24 h of aerobic incubation at 37°C, but isolation is also possible after microaerophilic or anaerobic incubation if this is necessary due to examinations for other bacteria. If the sole aim of the examination is to isolate *Yersinia*, cultivation at 25°C is recommended, while 42°C is recommended if *Salmonella* is the main target.

Table 10.2 Recipe for MacConkey Agar[a]

Sterile, deionized water	1000 ml
Peptone	20.0 g
Lactose	10.0 g
Sodium chloride	5.0 g
Ox bile	106 ml
Neutral red	0.075 g
Agar	9.0 g

Note: pH is stabilized at 7.2.

[a] The agar is commercially available from Gibco (U.S.).

Samples for *Salmonella* should be propagated in a propagation broth if healthy animals are examined in routine investigations. Several such media are commercially available or can be prepared on site in the laboratory, for example, selenite broth[9] (Table 10.3). After a short incubation time an indicative and/or selective agar, such as BPLS agar (Table 10.4), is streaked from the broth. As propagated *Salmonella* bacteria do not survive for too long in a broth, the incubation time of the propagation broth should be kept below 24 h.

Table 10.3 Recipe for Selenite Broth[a]

Sterile, deionized water	1000 ml
Peptone	5.0 g
Lactose	4.0 g
Sodium hydrogen selenite	6.0 g
Disodium hydrogen phosphate, 12H$_2$O	10.0 g

Note: pH is stabilized at 7.0.

[a] Samples are inoculated directly into 7 ml of the broth and incubated at 37°C aerobically for 12 to 24 h, after which an indicative agar is streaked from the broth. The broth is commercially available from Gibco (U.S.).

Table 10.4 Recipe for Brilliant Green Lactose
Sucrose (BPLS) Agar[a]

Sterile, deionized water	1000 ml
Peptone 140 (Pancreatic digest of casein)	7.0 g
Peptone 100 (Pancreatic digest of animal tissue)	7.0 g
Yeast extract	5.0 g
Lactose	10.0 g
Sucrose	10.0 g
Sodium chloride	5.0 g
Phenol red	0.04 g
Brilliant green	0.004 g
Agar	14.0 g

Note: Should only be boiled to dissolve the components. pH is stabilized at 6.9.

[a] Gram-positive and most Gram-negative bacteria, except for *Salmonella*, are inhibited. *Salmonella* grows as red colonies, while lactose/sucrose fermenters, such as *Proteus*, which will occasionally grow on this agar, occur as yellow colonies. Also, *Pseudomonas* occasionally grows with red colonies, but may easily be differentiated if a colony is subjected to a test for cytochrome oxidase. The agar is commercially available from Gibco (U.S.) or Merck (Germany), for example.

Main sample sites for Enterobacteriaceae are the cecum, feces, and genitals, although they may be isolated from several other sites as well.

10.1.2.2 Identification
After growing pure cultures of all isolates, each isolate should be subjected to microscopy, tests for Gram properties, oxidase, catalase, motility, and ability to grow in KCN. If the isolate is a member of Enterobacteriaceae identification may be performed according to Tables 10.5 to 10.8. However, identification of Enterobacteriaceae is generally much easier using a commercial test kit rather than in-house tests. Several commercial kits for Enterobacteriaceae are available, e.g., API 20E (bioMérieux, France), Enterotube II (Becton Dickinson, France), or Micro-ID (Organon Teknika, U.S.). It should be kept in mind that these kits were not developed for laboratory animal use, and therefore the identification procedure will not reveal all rodent-specific bacteria. The API 20E identification codes for *C. rodentium* are 1604513, 1404513, 1204513, 1104513, or 1004513.

If only specific types of Enterobacteriaceae are searched for, a screening test should be included to limit the number of isolates that must be fully identified. Tests for motility and KCN resistance as proposed by Tables 10.5 to 10.8 are valuable tools and may be combined with cultivation on a lactose-sucrose agar in routine examination. For example, in the FELASA guidelines for health monitoring of laboratory animals,[10] only *C. rodentium* and *Salmonella* are mandatory agents and both are lactose-sucrose negative, while this

Table 10.5   KCN-Resistant, Motile *Enterobacteriaceae* spp. Found in Rodents and Rabbits

	Citrobacter		Enterobacter					Hafnia alvei	Klyuvera	Morganella morganii	Proteus mirabilis	Serratia	
	freundii	rodentium	aerogenes	agglomerans	cloacae	gergoviae	sakazakii					liquefaciens	marquescens
KCN (growth)	+	+	+	d	+	d	+	+	+	+	+	+	+
Motility (37%)	+	+	+	d	+	+	+	d	+	+	+	+	+
β-Galactosidase	d	+	+	+	+	+	+	d	+	−	−	+	+
Arginine decarb.	d	−	−	−	d	d	+	−	d	−	−	−	−
Lysine decarb.	−	−	+	−	−	+	−	+	d	−	−	d	+
Ornithine decarb.	d	+	+	d	d	d	d	+	+	+	d	+	+
Citrate	d	d	d	d	d	d	+	d	d	−	d	d	+
H2S	d	−	−	d	−	−	−	−	−	−	d	−	−
Urease	d	−	−	d	−	+	−	−	−	+	d	−	d
Tryptophane deam.	−	−	−	d	−	−	−	−	−	+	+	−	−
Indole	d	−	−	d	−	+	d	d	d	+	−	−	−
VP	−	−	d	d	d	−	d	−	−	−	d	d	d
Gelatine liquefaction	−	+	+	+	+	+	+	d	+	+	−	d	d
Glucose (gas)	d	−	+	d	+	d	+	d	+	−	d	d	d
Lactose	d	+	+	d	+	+	+	+	+	−	+	+	+
Sucrose	d	−	+	+	+	+	+	d	+	d	−	−	−
Mannitol	+	+	+	d	+	d	+	+	+	+	+	+	+
Inositol	−	−	−	d	d	−	d	−	−	−	−	+	+
Sorbitol	+	+	+	−	d	−	−	−	−	−	−	+	d

Rhamnose	+	+	+	d	d	+	+	d	+	–	–	–	–
Sucrose	d	–	+	d	+	+	+	–	–	–	–	+	+
Melibiose	d	–	+	d	d	+	+	–	+	–	–	d	d
Amygdalin	d	+	+	+	+	+	+	d	+	–	–	+	+
Arabinose	+	+	+	+	+	+	+	d	+	–	–	d	d

Table 10.6 KCN-Sensitive, Motile *Enterobacteriaceae* spp. Found in Rodents and Rabbits

	Citrobacter	*Edwardsiella*	*Enterobacter*		*Escherichia*	*Salmonella*
	diversus	*tarda*	*agglomerans*	*gergoviae*	*coli* [a]	
KCN (growth)	–	–	d	d	–	–
Motility (37%)	+	+	d	+	+	+
β-Galactosidase	+	–	+	+	+	–
Arginine decarb.	d	–	–	–	–	d
Lysine decarb.	–	+	–	d	+	+
Ornithine decarb.	+	+	–	+	d	+
Citrate	+	–	d	d	–	d
H_2S	–	+	–	–	–	d
Urease	d	–	d	+	–	–
Tryptophane deam.	–	–	–	–	–	–
Indole	+	+	d	–	+	–
VP	–	–	d	+	–	–
Gelatine liquefaction	–	–	d	–	–	–

continued

Table 10.6 (continued) KCN-Sensitive, Motile *Enterobacteriaceae* spp. Found in Rodents and Rabbits

	Citrobacter diversus	Edwardsiella tarda	Enterobacter agglomerans	Enterobacter gergoviae	Escherichia coli[a]	Salmonella
Glucose (gas)	+	+	+	+	+	+
Lactose	d	–	d	d	+	–
Sucrose	d	–	d	+	d	–
Mannitol	+	–	+	+	+	+
Inositol	–	–	d	d	–	d
Sorbitol	+	–	d	–	–	+
Rhamnose	+	–	d	+	d	+
Sucrose	d	–	d	+	d	–
Melibiose	–	–	d	+	d	d
Amygdalin	+	–	+	+	d	–
Arabinose	+	–	+	+	+	+

[a] Inactive species of *E. coli* may occur.

Table 10.7 KCN-Resistant, Nonmotile *Enterobacteriaceae* spp.
Found in Rodents and Rabbits

	Providencia	*Klebsiella*			
	rettgeri	*orthinolytica*	*oxytoca*	*planticola*	*pneumoniae*
KCN (growth)	+	+	+	+	+
Motility (37%)	–	–	–	–	–
β-Galactosidase	–	+	+	+	+
Arginine decarb.	–	–	–	–	–
Lysine decarb.	–	+	d	+	d
Ornithine decarb.	–	+	–	–	–
Citrate	d	+	+	+	d
H₂S	–	–	–	–	–
Urease	+	d	d	+	d
Tryptophane deam.	+	–	–	–	
Indole	+	+	+	d	–
VP	–	d	d	+	d
Methyl red test	+	+	d	+	–
Gelatine liquefaction	–	–	–	–	–
Glucose (gas)	–	+	+	+	+
Lactose	d	+	+	+	+
Sucrose	–	+	+	+	+
Mannitol	d	+	+	+	+
Inositol	d	+	+	+	d
Sorbitol	–	+	+	+	+
Rhamnose	d	+	+	+	+
Sucrose	d	+	+	+	+
Melibiose	–	+	+	+	+
Amygdalin	d	+	+	+	+
Arabinose	–	+	+	+	+

Table 10.8 KCN-Sensitive, Nonmotile *Enterobacteriaceae* spp.
Found in Rodents and Rabbits

	Enterobacter agglomerans	*Yersinia pseudotuberculosis*
KCN (growth)	d	–
Motility (37%)	d	–
β-Galactosidase	+	d
Arginine decarb.	–	–
Lysine decarb.	–	–
Ornithine decarb.	–	–
Citrate	d	–
H₂S	–	–

continued

Table 10.8 (continued)   KCN-Sensitive, Nonmotile
Enterobacteriaceae spp. Found in Rodents and Rabbits

	Enterobacter agglomerans	*Yersinia pseudotuberculosis*
Urease	d	+
Tryptophane deam.	–	–
Indole	d	–
VP	d	–
Gelatine liquefaction	d	–
Glucose (gas)	+	–
Lactose	d	–
Sucrose	d	+
Mannitol	+	+
Inositol	d	–
Sorbitol	d	–
Rhamnose	d	+
Sucrose	d	–
Melibiose	d	d
Amygdalin	+	d
Arabinose	+	d

rather seldom occurs for *E. coli*, which is the most common species of Enterobacteriaceae found in rats and mice.

Detection of antibodies in serum samples is not a common tool for detection of Enterobacteriaceae in rodent colonies.

Full identification of *Salmonella* includes determination of O and H antigens, which can be done by slide agglutination using polyvalent sera. In most countries reference centers able to group *Salmonella* serologically are run on a national basis.

Salmonella may be detected by polymerase chain reaction (PCR). Nonradioactive-labeled probes may be used for identifying and visualizing the amplified products.

The sensitivities of colorimetric assays using peroxidase and alkaline phosphatase are similar to those obtained with an ethidium bromide-stained agarose gel, as both procedures allow the detection of as few as 50 *Salmonella*.[11] A short cultivation procedure, e.g., propagation in selenite broth, combined with a *Salmonella*-specific PCR-hybridization assay, may be applied to specifically identify *Salmonella* serovars from feces samples. Such an assay may detect as few as nine colony-forming units of *Salmonella* organisms in pure culture and as little as 300 fg of purified chromosomal DNA.[12]

10.1.3 Safety

The U.S. CDC recommends Biosafety Level-2 practices, containment equipment, and facilities for activities with cultures of, or clinical materials potentially infected with, *Salmonella.* Animal Biosafety Level-2 practices, containment equipment, and facilities are recommended for activities with experimentally or naturally infected animals.[13]

10.2 Pasteurellaceae

Pasteurellaceae are Gram-negative, mostly oxidase-positive, nonmotile, catalase-positive, pasteurellaform bacteria. No rodent species grow on MacConkey agar. *Pasteurella in sensu stricto* is found mostly in rabbits, while rodents harbor various sorts of *Haemophilus* and *Actinobacillus*, some of the latter of which, however, are traditionally designated as *P. pneumotropica.*

10.2.1 Pasteurella

10.2.1.1 Characteristics of infection

One agent, *P. multocida,* is of importance. It is a facultative pathogen of rabbits, and it may, in principle, infect rodents, which is uncommon. In Europe the agent seems to be less common, while it still remains a problem in conventional rabbitries in the U.S., where surveys have revealed the infection in up to 70% of conventional rabbits. Infection is mostly subclinical, and epizootic disease appears to be due largely to environmental and host-related factors, e.g., environmental changes or experimental procedures. Respiratory disease occurs as "snuffles," i.e., chronic and in certain cases atrophic rhinitis,[14] otitis media,[15] and pneumonia, which may vary from chronic purulent bronchopneumonia (enzootic pneumonia) to acute fibrinous pneumonia.[16] Respiratory problems occur most often in the spring and fall, and are lowest in the summer. Conjunctivitis, abscesses, and acute septicemias have also been described, and the agent also may be found in the genitals. Direct contact with animals shedding *P. multocida* from nasal or vaginal secretions is considered the chief means of spread. Suckling rabbits may be infected with *P. multocida* from carrier does within the first week of life. Individual animals are often chronically infected by 6 weeks of age.[17] The infection does not seem to spread between rabbits that are not in close contact with one another.[18-20] A barrier system is an efficient way of keeping rabbits free of the infection.[21] Transmission from other species, e.g., pigs and cattle, may occur.[22]

It should be noted that *"Pasteurella"* spp., such as *P. pneumotropica* found in various rodents, are no longer classified as *Pasteurella* but as *Actinobacillus* (see below).

10.2.1.2 Characteristics of the agent

10.2.1.2.1 Morphology. P. multocida are small, coccoid, Gram-negative rods (Figure 4.1[8]). They often occur as chains or pairs and staining often is bipolar. Colonies on blood agar are nonhemolytic and 1 to 3 mm after

24 h. Virulent strains are smooth or mucoid, while rough colonies often indicate a low virulence.[23]

10.2.1.2.2 *Cultivation.*   Samples should be inoculated directly on blood or chocolate agar. After 24 to 48 h of aerobic cultivation at 37°C colonies should be clearly visible.

10.2.1.2.3 *Isolation sites.*   From diseased animals the affected organ should be sampled. Additionally, blood may be sampled if septicemia is suspected. From healthy animals the most appropriate sampling sites are the nose, trachea, lungs, and genitals.

10.2.1.2.4 *Differentiation and identification.*   Biochemical characteristics for differentiating *P. multocida* from related species are given in Table 10.9. Three subspecies — subsp. *multocida,* subsp. *septica,* and subsp. *gallicidae* — have been defined. These differ as indicated in Table 10.10. The commercial kit API 20NE (bioMérieux, France) may be used for identification of *P. multocida.*

PCR amplification performed directly on bacterial colonies or cultures represents an extremely rapid, sensitive method for identification of *P. multocida*[24] and for dividing these into different subtypes.[25,26]

10.2.1.2.5 *Serology.*   ELISA[27] or dot blotting[28] may be used to diagnose infection with *P. multocida* in rabbit colonies.

Table 10.9 Differentiation of Non-X or V-Factor-Dependent *Pasteurellaceae* Found in Laboratory Rodents and Rabbits

	Pasteurella multocida	*"Pasteurella" pneumotropica*		*Actinobacillus muris*
		Jawetz	Heyl	
Lactose	−	+	+	−
Mannitol	+	−	−	+
D-Galactose	+·	+	+	−
Raffinose	−	+	−	+
Aesculin	−	−	−	+
Urease	−	+	+	−
Ornithine decarb.	+	+	+	−
Indole	d	+	+	−
L-Arabinose	d	−	+	−
Trehalose	d	+	+[a]	+
D-Mannose	+	+	+	+
Xylose	d	+	+	−
Melibiose	d	+	−	+

[a] Certain not clearly defined species from rats may be negative for trehalose.

10.2.2. Actinobacillus

Two species are of interest. *Actinobacillus muris* is a seldom isolated bacterium of rats and mice, while *"Pasteurella" pneumotropica* is an important rodent bacterium formerly classified as *Pasteurella*, which, after a reclassification in 1985, was classified as *Actinobacillus*.[29] Occasionally, rodent isolation of *"Pasteurella haemolytica,"* which has also been reclassified, is reported. However, this agent is found primarily in farm animals, and closer examination of so-called *P. haemolytica* from rodents usually identifies these as *P. pneumotropica*.

10.2.2.1 Characteristics of infection

P. pneumotropica was first described by Jawetz in 1948.[30] Until then the etiologic agent of Pasteurella-associated disease in rodents was diagnosed as *P. multocida*. Infections with *P. pneumotropica* are mostly latent. Pasteurellosis of rats has been described as either upper respiratory disease such as

Table 10.10 Differing Subspecies of *Pasteurella multocida*

	Pasteurella multocida		
	subsp. *multocida*	subsp. *septica*	subsp. *gallicidae*
Sorbitol	+	−	+
Dulcitol	−	−	+

rhinitis, sinuitis, otitis media, conjunctivitis, and opthalmitis, or pyogenic infections such as subcutaneous abscesses or mastitis.[31-33] However, currently it is generally accepted that *Pasteurella* as a pathogen in rodents is mainly secondary to a primary agent, such as *Mycoplasma pulmonis* or Sendai virus. Experimental pulmonary disease caused by *P. pneumotropica* and *M. pulmonis* in combination resembles naturally occurring chronic respiratory conditions more than experimental conditions produced by either one of these organisms.[34] Stress — including experimental stress or immunosuppression — may activate latent infections. The incidence of spontaneous deaths during inhalation anesthesia in rodents might be raised in *P. pneumotropica*-infected animals.

Transmission is mainly horizontal by droplets, but pseudovertical contamination may occur, as intrauterine infection may reach prevalences as high as 60 to 70%.[35-37] Therefore, puppies may be infected during birth or Cesarean section. One should, therefore, never introduce the uterus into the isolator during Cesarean section. Rats are most easily infected with *P. pneumotropica* strains of rat origin. The same specificity is observed in mice. Experimental cross-infection is possible.

The number of rodent colonies infected with *P. pneumotropica* is highest for conventional colonies, perhaps close to 100%, but many barrier-bred colonies of rats, mice, and gerbils also harbor one or both of these agents.

Carrier prevalences in infected colonies vary from 48 to 95%.[38–40] *P. pneumo-tropica* also is isolated from rabbits in rare cases.

A. *muris* may cause infertility due to abortion, metritis, and stillbirths in mice.[41] The infection rarely occurs, and when present it is mostly latent. This rodent agent was formerly identified as *Pasteurella ureae* or *Actinobacillus urea*, which, however, seems to be a specific human agent.[42]

10.2.2.2 Characteristics of the agent

10.2.2.2.1 Morphology.   The bacterial morphology is typical pasteurella form (Figure 4.1[8]). Colonies are small, 0.5 to 1 mm, white or yellow, with a certain smell (described by some as close to the smell of male mice, by others as the smell of ink or sperm).

10.2.2.2.2 Cultivation.   Chocolate agar should be used as the primary medium for the isolation of both *P. pneumotropica* and *A. muris*, although they will also grow on some types of blood agar. However, as there are certain types of blood that do not support growth, the safest procedure is to use chocolate agar (see Chapter 3). Cultivation for 24 h at 37°C will generally prove to be sufficient.

10.2.2.2.3 Isolation sites.   The vagina or prepuce often seems to be the most successful site of isolation for these agents, which also may be isolated from nasoturbinates, conjunctiva, nasolacrimal duct, trachea, and lungs. *P. pneumotropica* may also be found in the uterus, and it may during long periods be latent in the intestines from where it is difficult to isolate. Intestinal infection only seldom spreads to other organs.[43]

10.2.2.2.4 Differentiation and identification.   *P. pneumotropica* may be divided biochemically into two biotypes, type Jawetz and type Heyl. A third type, type Henriksen, isolated from humans, is now classified as *P. dagmatis* and has no importance in laboratory animal bacteriology. Full identification may be performed according to Table 10.9. It should be noted that *P. pneumotropica* is a rather slow fermenter and, therefore, long-term incubation, i.e., 48 h, may be necessary before safe conclusions should be drawn on the basis of fermentation results. Some species specificity seems to exist between variants of *P. pneumotropica*. Mouse-associated strains are always trehalose positive, while rat-associated strains are occasionally negative. Mouse strains often hemagglutinate human erythrocytes, which rat strains do not.[44,45] The commercial kit API 20NE (bioMérieux, France) is a valuable diagnostic tool for diagnosis of *P. pneumotropica* if the basic rules for the use of commercial kits (Chapter 4) are respected, the slow reactions of *P. pneumotropica* are noted, and the fact that *P. haemolytica* is not found in rodents is also kept in mind. Mannitol fermentation, ornithine decarboxylase, and indole reaction are traditionally used for differentiating between *P. pneumotropica* and *A. muris*. The latter is not included in the API system.

10.2.2.2.5 Serology. Antibodies against *P. pneumotropica* in sera from infected animals may be detected in an agglutination test, complement-fixation test, or ELISA.[46–48] These tests are valuable diagnostic tools, but most of them diagnose the presence of various members of Pasteurellaceae, and not solely *P. pneumotropica*.

10.2.2.2.6 Molecular biology. Infections with Pasteurellaceae may be diagnosed by PCR. A method described by Boot et al.[49] may be used on samples from the respiratory system to detect the presence of several members of Pasteurellaceae. The method seems to be more sensitive than cultivation. Another method described by Weigler et al.[50] may be useful for identification of isolates of *P. pneumotropica*.

10.2.3 Haemophilus

10.2.3.1 Characteristics of infection
Haemophilus spp. is a common finding in colonies of rats, mice, guinea pigs, and rabbits, even if these animals are barrier bred. One species from mice, *H. influenza murium*, is clearly defined,[51] while species isolated from rats are less characterized. The pathogenicity is low, although infection in rats has been associated with mild inflammatory cell infiltration and a light diffuse hyperemia in the lungs.[52] Prevalences in infected colonies are generally below 20%.[40]

10.2.3.2 Characteristics of the agent
Haemophilus are pasteurella forms (Figure 4.1[8]) varying from coccobacilli to more rod-like cells. If testing for cytochrome oxidase is performed as described by Kovács[53] (Table 4.5) and not as later described by Gaby and Hadley,[54] *Haemophilus* spp. are oxidase positive. Growth depends on either X-factor (protoporphyrin or protohemin), V-factor (nicotine amide dinucleotide, NAD), or both. Growth is most simply achieved by inoculation on chocolate agar. Lincomycin (5 μg/ml)[55] or clindamycin (2 μg/ml) may be added to prevent growth of Gram-positive bacteria. Incubation should be at 37°C, aerobic or microaerophilic, for 24 to 48 h. The main sampling sites are within the respiratory system and from the vagina or the prepuce.

Dependency on X- or V-factor is tested by the use of discs containing the factors. Such discs are commercially available from Rosco (Denmark) for example. Three types of discs are used: one with the X-factor, one with the V-factor, and one with both. The test is performed as other disc methods (see Chapter 4) using Mueller Hinton agar. Isolates growing close to only one of the single-factor discs *and* the X-V factor disc are considered dependent on that factor, while isolates growing only close to the disc containing both factors and eventually between the X-V-factor discs are considered dependent on both factors.

For the test of biochemical characteristics the X- and V-factor must be added to the media. *H. influenza murium* may be identified according to Table

Table 10.11   Differentiation of X- or V-Factor-Dependent
Pasteurellaceae Found in Laboratory Rodents and Rabbits

	Haemophilus influenza murium	Other spp.
X-factor dependent (hem)	+	−
V-factor dependent (NAD)	d	+
Esculin	−	−
Urease	−	−
Arabinose	−	−
Mannitol	−	d
Phosphatase	−	d
Indole	−	d
Xylose	−	d
Trehalose	+	d
Melibiose	−	d
Raffinose	−	d
Ribose	+	d

10.11. Identification tests may most easily be carried out using the commercial kit API NH (bioMérieux, France), although the computer software or the profiles supplied with the kit should not be expected to give any clear result for rodent species. API NH codes for some rat *Haemophilus* spp. are given in Table 10.12.

ELISA and IFA may be attempted for colony diagnosis. Although cross-reactions between different strains of *Haemophilus* and *Pasteurella pneumotropica* may occur, ELISA for *P. pneumotropica* should not be relied on as a secure tool for revealing infection with *Haemophilus*. PCR is applicable for diagnosis on samples from the pharynx, trachea, and lungs.[50]

10.3 Streptobacillus moniliformis

10.3.1 Characteristics of infection

Streptobacillus moniliformis is the sole species in this genus. It may be isolated from mice, rats, and guinea pigs, and it is transmissible to humans, in which it causes *rat bite fever*, a purulent wound infection developing into petechial exanthema, polyarthritis, fever, and death. Currently it is not a common zoonotic condition in laboratory animal facilities, but fatal cases have occurred recently in the Western world after contact with wild rats. It has also recently been isolated from barrier-maintained mice,[56] but this is to be regarded as the exception rather than the rule. Also, *Spirillum minus* (Chapter 13) may be the cause of rat bite fever.

The laboratory animal most susceptible to the development of disease is the mouse, in which cases begin as swelling of the cervical lymph nodes,

Table 10.12 Codes for Some Isolates of *Haemophilus* from Rats, Guinea Pigs, and Rabbits as Determined by the Commercial Kit API NH (bioMérieux, France)[61,62]

API NH code	Animal origin	X-factor dependent	V-factor dependent	Glucose	Fructose	Maltose	Sucrose	Ornithine decarboxylase	Urease	Lipase	Alkaline phosphatase	β-Galactosidase	Prolinearyl amidase	γ-glutamyl transferase	Indole
7766	Rat, Guinea pig	–	+	+	+	+	+	+	+	–	+	+	–	+	+
7520	Rat	–	+	+	+	+	+	–	+	–	+	–	–	–	–
7760	Rat	–	+	+	+	+	+	–	+	–	+	+	–	–	–
7320	Rat	–	+	+	+	+	+	+	–	–	+	–	–	–	–
7720	Rat, Guinea pig	–	+	+	+	+	+	+	+	–	+	–	–	–	–
7360	Rat, Guinea pig	–	+	+	+	+	+	+	–	–	+	+	–	–	–
7162	Guinea pig	–	+	+	+	+	+	–	–	–	+	+	–	+	–
7122	Rabbit	–	+	+	+	+	+	–	–	–	+	–	–	+	–

which may turn into fatal septicemia. Chronic cases are characterized by arthritis in the distal parts of the legs and the tail. Abscesses[57] and abortions[58] may occur. Genetic factors seem to be rather essential for the susceptibility to the development of disease and perhaps for susceptibility to infection. Specifically, C57BL/6 mice seem to be highly susceptible, and the agent is difficult to isolate from nonsusceptible strains kept in the same unit as an infected, susceptible strain.[56]

In guinea pigs the agent causes local abscesses, which do not spread. In rats no clinical signs should be expected.

10.3.2 Characteristics of the agent

10.3.2.1 Morphology
Cells of *S. moniliformis* are highly pleomorphic. Most cells are less than 1 μm wide and less than 5 μm long, but they may be up to 150 μm long. Branching is not observed. The colonies are 1 to 2 mm, round, grayish, smooth, and glistening. Also, fried-egg-type colonies may be observed.

10.3.2.2 Cultivation
Isolation media should be highly enriched, e.g., 20% serum should be added. Both agar and broth should be included in screening procedures. The blood or chocolate agars described in Chapter 3 may be applied, while agars containing polyanetholesulfonate seem to be inhibitory.[59] The media are incubated microaerophilically (8% CO_2) for 1 to 6 days. Agars should be inspected daily for growth, while broths should be inspected daily for so-called "puff balls."

10.3.2.3 Isolation sites
These bacteria are extremely difficult to isolate from healthy animals. The nose, trachea, and genitals are the most appropriate sites in healthy animals, while joint fluid and blood should be included if disease is observed.

10.3.2.4 Differentiation and identification
S. moniliformis is the only catalase-negative, Gram-negative, facultatively anaerobic rod in rodent bacteriology, and the microscopic morphology also should be helpful in the identification process. It is oxidase negative and does not grow on MacConkey agar. Furthermore, the profile given in Table 10.13 may be used.

10.3.2.5 Serology
ELISA to measure *S. moniliformis* antibodies in mice and rats has been developed. Different *S. moniliformis* strains originating from cases of rat bite fever in man and from various rodent species show considerable

Table 10.13 Biochemical
Characteristics
of *Streptobacillus moniliformis*

D-Glucose	+
Lactose	−
Maltose	+
D-Mannitol	−
Sucrose	−
D-Xylose	−
Citrate	−
Indole	−
Nitrate reduction	−

serological relationship. ELISA reveals more contaminated animals than does cultivation.[60]

10.4 Vibrionaceae

The only genus in this family to be isolated frequently from rodents is *Aeromonas*. They may be isolated from the cecum, genitals, and respiratory organs. Prevalences are low. They are often part of a mixed secondary flora in relation to wound infection, respiratory disease, etc., but have little importance as primary pathogens.

They share some common characteristics with *Vibrio* spp. and therefore differentiation in principle is necessary, although *Vibrio* is not a common finding in laboratory animals. This can be done by testing for sensitivity to O/129 (2,4-diamino-6,7-diisopropylpteridine), which is done by using a 150-µg disk in the agar diffusion inhibition assay (see Chapter 4). *Aeromonas* is resistant, while most *Vibrio* are sensitive.

Aeromonas found in rodents are either *A. hydrophila*, *A. caviae*, or *A. veronii* subsp. *sobria*, which may be differentiated according to Table 10.14. The commercial kit API 20NE (bioMérieux, France) is a helpful tool for identification on a genus level, but it is not very usable for differentiation of species.

Table 10.14 Differentiation of *Aeromonas* spp. Found in Rodents

		Aeromonas	
	caviae	*hydrophila*	*veronii* subsp. *sobria*
Gas from glucose	−	+	+
Aesculin	−	−	+

References

1. Ocholi, R. A., Chima, J. C., Uche, E.M., and Oyetunde, I. L., An epizootic infection of *Citrobacter freundii* in a guinea pig colony: short communication, *Lab. Anim.*, 22, 335, 1988.
2. Barthold, S. W., Coleman, G. L., Bhatt, P. N., Osbaliston, G. W., and Jonas, A. M., The etiology of murine colonic hyperplasia, *Lab. Anim. Sci.*, 26, 889, 1976.
3. Brennan, P. C., Fritz, T.E., Flynn, R. J., and Poole, C. M., *Citrobacter freundii* associated with diarrhea in laboratory mice, *Lab. Anim. Care*, 15, 266, 1965.
4. Ediger, R. D., Kovatch, R. M., and Rabstein, M. M., Colitis in mice with a high incidence of rectal prolapse, *Lab. Anim. Sci.*, 24, 488, 1974.
5. Barthold, S. W. and Jonas, A. M., Morphogenesis of early 1,2-dimethylhydra-zine-induced lesions and latent period reduction of colon carcinogenesis in mice by a variant of *Citrobacter freundii*, *Cancer Res.*, 37, 4352, 1977.
6. Okewole, P. A., Uche, E. M., Oyetunde, I. L., Odeyemi, P. S., and Dawaul, P. B., Uterine involvement in guinea pig salmonellosis, *Lab. Anim.*, 23, 275, 1989.
7. Hansen, A. K., Antibiotic treatment of nude rats and its impact on the aerobic bacterial flora, *Lab. Anim.*, 29(1), 37, 1995.
8. MacConkey, A. T., Bile salt media and their advantages in some bacteriological examinations, *J. Hyg. (Camb.)*, 8, 322, 1908.
9. Leifson, E., New selenite enrichment media for the isolation of typhoid and paratyphoid (*Salmonella*), *Amer. J. Hyg.*, 24, 423, 1936.
10. Kraft, V., Deeney, A. A., Blanchet, H. M., Boot, R., Hansen, A. K., Hem, A., van Herck, H., Kunstyr, I., Milite, G., Needham, J. R., Nicklas, W., Perrot, A., Rehbinder, C., Richard, Y., and de Vroy, G., Recommendations for the health monitoring of mouse, rat, hamster, guinea pig and rabbit breeding colonies, *Lab. Anim.*, 28, 1, 1994.
11. Soumet, C., Ermel, G., Boutin, P., Boscher, E., and Colin, P., Chemiluminescent and colorimetric enzymatic assays for the detection of PCR-amplified *Salmonella* sp. products in microplates, *Biotechniques*, 19(5), 792, 1995.
12. Stone, G. G., Oberst, R. D., Hays, M. P., McVey, S., and Chengappa, M. M., Detection of Salmonella serovars from clinical samples by enrichment broth cultivation-PCR procedure, *J. Clin. Microbiol.*, 32(7), 1742, 1994.
13. Office of Health and Safety, Centers for Disease Control and Prevention, Laboratory Biosafety Level Criteria, http://www.cdc.gov/od/ohs/bios-fty/bmbl/section3.htm.
14. DiGiacomo, R. F., Deeb, B. J., Giddens, W. E., Jr., Bernard, B. L., and Chen-gappa, M. M., Atrophic rhinitis in New Zealand rabbits infected with *Pasteurella multocida*, *Amer. J. Vet. Res.*, 50, 1460, 1987.
15. Flatt, R. E., Deyoung, D. W., and Hogle, R. M., Suppurative otitis media in the rabbit: prevalence, pathology, and microbiology, *Lab. Anim. Sci.*, 27, 343, 1977.
16. Percy, D. H., Pathology of Laboratory Animals, The Rabbit (*Oryctolagus cuniculus*), http://www.afip.org/vetpath/POLA/POLA96/rabbit.cvp, 1995.
17. Holmes, H. T., Patton, N. M., and Cheeke, P. R., The occurrence of *Pasteurella multocida* in newborn and weanling rabbits, *J. Appl. Rabbit Res.*, 6, 125, 1983.
18. Percy, D. H., Bhasin, J. L., and Rosendal, S., Experimental pneumonia in rabbits inoculated with strains of *P. multocida*, *Can. J. Vet. Res.*, 50, 36, 1986.

19. Lelkes, L. and Corbett, M. J., A preliminary study of the transmission of *P. multocida* in rabbits, *J. Appl. Rabbit Res.*, 6, 125, 1983.
20. DiGiacomo, R. F., Jones, C. D., and Wathes, C. M., Transmission of *Pasteurella multocida* in rabbits, *Lab. Anim. Sci.*, 37, 621, 1987.
21. Scharf, R. A., Monteleone, S. A., and Stark, D. M., A modified barrier system for maintenance of *Pasteurella*-free rabbits, *Lab. Anim. Sci.*, 31, 513, 1981.
22. Al-Lebban, Z. S., Corbeil, L. B., and Coles, E. H., Rabbit pasteurellosis: induced disease and vaccination, *Am. J. Vet. Res.*, 49, 312, 1988.
23. Ohder, H. and Wullenweber, M., *Pasteurella* sp., in *Diagnostic Microbiology for Laboratory Animals*, Kunstyr, I., Ed., Gustav Fischer, New York, 1992, 87.
24. Townsend, K. M., Frost, A. J., Lee, C. W., Papadimitriou, J. M., and Dawkins, H. J., Development of PCR assays for species- and type-specific identification of *Pasteurella multocida* isolates, *J. Clin. Microbiol.*, 36(4), 1096, 1998.
25. Zucker, B., Kruger, M., and Horsch, F., Differentiation of *Pasteurella multocida* subspecies *multocida* isolates from the respiratory system of pigs by using polymerase chain reaction fingerprinting technique, *Zentralbl. Veterinarmed.* [B], 43(10), 585, 1996.
26. Chaslus-Dancla, E., Lesage-Decauses, M. C., Leroy-Setrin, S., Martel, J. L., Coudert, P., and Lafont, J. P., Validation of random amplified polymorphic DNA assays by ribotyping as tools for epidemiological surveys of *Pasteurella* from animals, *Vet. Microbiol.*, 52(1–2), 91, 1996.
27. Lukas, V.S., Ringler, D. H., Chrisp, C. E., and Rush, H. G., An ELISA to detect serum IgG to *Pasteurella multocida* in naturally and experimentally infected rabbits, *Lab. Anim. Sci.*, 37, 60, 1987.
28. Manning, P. J., Brackee, G., Naasz, M. A., DeLong, D., and Leary, S. L., A dot-immunobinding assay for the serodiagnosis of *Pasteurella multocida* infection in laboratory rabbits, *Lab. Anim. Sci.*, 37, 615, 1987.
29. Mutters, R., Ihm, P., Pohl, S., Frederiksen, W., and Mannheim, W., Reclassification of the genus *Pasteurella* Trevisan 1887 on the basis of DNA homology with proposals for the new species *Pasteurella dagmatis*, *Pasteurella canis*, *Pasteurella stomatis* and *Pasteurella langaa*, *Int. J. System. Bacteriol.*, 35, 309, 1985.
30. Jawetz, E., A latent pneumotropic *Pasteurella* of laboratory animals, *Proc. Soc. Exp. Biol. Med.*, 68, 46, 1948.
31. Wheater, D. F. W., The bacterial flora of an SPF colony of mice, rats and guinea pigs, in *Husbandry of Laboratory Animals*, Conlaty, M. L., Ed., Academic Press, New York, 1967, 343.
32. van der Schaff, A., Mullink, J. W. M. A., Nikkels, R. J., and Goudswaard, J., *Pasteurella pneumotropica* as a causal microorganism of multiple subcutaneous abscesses in a colony of Wistar rats, *Z. Versuchstierkd.*, 12, 356, 1970.
33. Young, C. and Hill, A., Conjunctivitis in a colony of rats, *Lab. Anim.*, 8, 301, 1974.
34. Brennan, P. C., Fritz, T. E., and Flynn, R. J., The role of *Pasteurella pneumotropica* and *Mycoplasma pulmonis* in murine pneumonia, *J. Bacteriol.*, 97, 337, 1969.
35. Graham, W. R., Recovery of pleuropneumonia-like organism (PPLO) from the genitalia of the female albino rat, *Lab. Anim. Care*, 13, 719, 1963.
36. Casillo, S. and Blackmore, D. K., Uterine infection caused by bacteria and mycoplasma in mice and rats, *J. Comp. Pathol.*, 82, 482, 1972.
37. Larsen, B., Markovetz, A.J., and Galask, R.P., The bacterial flora of the female rat genital tract, *Proc. Soc. Exp. Biol. Med.*, 151, 571, 1976.

38. Nakagawa, M., Saito, M., Suzuki, E., Nakayama, K., Matsubara, J., and Muto, T., Ten-years-long survey on pathogen status of mouse and rat breeding colonies, *Exp. Anim.*, 33(1), 11, 1984.
39. Hansen, A. K., The aerobic bacterial flora of laboratory rats from a Danish breeding centre, *Scand. J. Lab. Anim. Sci.*, 19(2), 59, 1992.
40. Besch-Wiliford, C. and Wagner, J.E., *Pasteurella pneumotropica*, in *Manual of Microbiologic Monitoring of Laboratory Animals*, Allen, A. M. and Nomura, T., Eds., U.S. Department of Health and Human Services, II.E.1–II.E.4, Washington, D.C., 1986.
41. Ackerman, J. I. and Fox, J. G., Isolation of *Pasteurella ureae* from reproductive tracts of congenic mice, *J. Clin. Microbiol.* 13(6), 1049, 1981.
42. Mutters, R., Frederiksen, W., and Mannheim, W., Lack of evidence for the occurrence of *Pasteurella ureae* in rodents, *Vet. Microbiol.*, 9, 83, 1984.
43. Moore, T. D., Allen, A. M., and Ganaway, J. R., Latent *Pasteurella pneumotropica* infection of the intestine of gnotobiotic and barrier-held rats, *Lab. Anim. Sci.*, 5, 657, 1973.
44. Boot, R., Epizootiological studies of *P. pneumotropica*, *GV-SOLAS Wiss. Tagung*, 1988.
45. Nicklas, W. and Mauter, P., Biochemische Untersuchungen an Pasteurella pneumotropica Isolaten von Versuchsgnagern, *GV-SOLAS Wiss. Tagung*, 1989.
46. Hoag, W.G., Wetmore, P.W., Rogers, J., and Meier, H., A study of latent *Pasteurella* infection in a mouse colony, *J. Infect. Dis.*, 11, 135, 1962.
47. Weisbroth, S. H., Scher, S., and Boman, I., *Pasteurella pneumotropica* abscess syndrome in a mouse colony, *J. Am. Vet. Med. Assoc.*, 155, 1206, 1969.
48. Manning, P.J., DeLong, D., Gunther, R., and Swanson, D., An enzyme-linked immunosorbent assay for detection of chronic subclinical *Pasteurella pneumotropica* infection in mice, *Lab. Anim. Sci.*, 41(2), 162, 1991.
49. Boot, F., Kirschnek, S., Nicklas, W., Wyss, S. K., and Homberger, F., Detection of Pasteurellaceae in rodents by polymerase chain reaction, *Lab. Anim. Sci.*, 48(5), 542, 1998.
50. Weigler, B. J., Wiltron, L. A., Hancock, S. I., Thigpen, J. P., Goelz, M. F., and Forsythe, D. B., Further evaluation of a diagnostic polymerase chain reaction assay for *Pasteurella pneumotropica*, *Lab. Anim. Sci.*, 48(2), 193, 1998.
51. Csukas, Z., Reisolation and characterization of *Haemophilus influenzae-murium*, *Acta. Microbiol. Acad. Sci. Hung.*, 23(1), 89, 1976.
52. Nicklas, W., *Haemophilus* infection in a colony of laboratory rats, *J. Clin. Microbiol.*, 27(7), 1636, 1989.
53. Kovács, N., Identification of *Pseudomonas pyocyanea* by the oxidase reaction, *Nature*, 178, 103, 1956.
54. Gaby, W. L. and Hadley, C., Practical laboratory tests for *Pseudomonas aeruginosa*, *J. Bacteriol.* 74, 356, 1957.
55. Csukas, Z., New selective medium for the isolation of *Haemophilus* species, *Acta Microbiol. Acad. Sci. Hung.*, 27(2), 141, 1980.
56. Wullenweber, M., Kaspareit-Rittinghausen, J., and Farouq, M., *Streptobacillus moniliformis* epizootic in barrier-maintained C57BL/6J mice and susceptibility to infection of different strains of mice, *Lab. Anim. Sci.*, 40, 608, 1990.
57. Kaspareit-Rittinghausen, J., Wullenweber, M., Deerberg, F., and Farouq, M., Pathologische Veränderungen bei *Streptobacillus moniliformis* Infektion von C57BL/6J Mäusen, *Berl. Münch. Tierärzl. Wschr.*, 103, 84, 1990.

58. Sawicki, L., Bruce, H. M., and Andrewes, C. H., *Streptobacillus moniliformis* as a probable cause of arrested pregnancy and abortion in laboratory mice, *Br. J. Exp. Pathol.*, 43, 194, 1962.
59. Lambe, D.W., Jr., McPhedran, M.W., Mertz, J.A., and Stewart, P., *Streptobacillus moniliformis* isolated from a case of Haverhill fever: biochemical characterization and inhibitory effect of polyanethol sulphonate, *Am. J. Clin. Pathol.*, 60, 854, 1973.
60. Boot, R., Bakker, H., Thuis, H., Veenema, J. L., and De Hoog, H., An enzyme-linked immunosorbent assay (ELISA) for monitoring rodent colonies for *Streptobacillus moniliformis* antibodies, *Lab. Anim.*, 27(4), 350, 1993.
61. Boot, R., Thuis, H. C. W., and Veenema, J. L., Serological relationship of some V-factor dependent Pasteurellaceae (*Haemophilus* spp.) from rats, *J. Exp. Anim. Sci.*, 38, 147, 1996/1997.
62. Boot, R., Thuis, H. C. W., and Veenema, J. L., Serological relationship of some V-factor dependent Pasteurellaceae (*Haemophilus* spp.) from guinea pigs and rabbits, *Lab Anim.*, 33, 91, 1999.

chapter eleven

Obligate aerobic, Gram-negative rods

Contents

11.1 Motile, aerobic, Gram-negative rods..192
    11.1.1 *Pseudomonas* 192
        11.1.1.1 Characteristics of infection...192
        11.1.1.2 Characteristics of the agent..194
            11.1.1.2.1 Morphology..194
            11.1.1.2.2 Cultivation..195
            11.1.1.2.3 Isolation sites...195
            11.1.1.2.4 Differentiation and identification...............195
            11.1.1.2.5 Serology..195
    11.1.2 *Xanthomonas, Sphingomonas, Chryseomonas, Agrobacterium,*
        and *Burkholderia* 195
    11.1.3 *Bordetella* 197
        11.1.3.1 Characteristics of infection...197
        11.1.3.2 Characteristics of the agent..197
            11.1.3.2.1 Morphology..197
            11.1.3.2.2 Cultivation..197
            11.1.3.2.3 Isolation sites...197
            11.1.3.2.4 Differentiation and identification...............197
            11.1.3.2.5 Serology..197
11.2 Nonmotile, aerobic, Gram-negative rods that can
    easily be cultivated...198
    11.2.1 *Acinetobacter* 198
    11.2.2 *Flavobacterium* 198
    11.2.3 *Weeksella*...198
11.3 Nonmotile, aerobic, Gram-negative rods that
    are difficult to cultivate..198

    11.3.1 *Francisella*..200
            11.3.1.1 Characteristics of infection..............................200
            11.3.1.2 Characteristics of the agent............................200
            11.3.1.3 Safety 200
References...201

This group is in no way to be regarded as a uniform group of bacteria. Basic tests, including tests for motility and oxidase activity, should allow the allocation of isolates to some more well-defined genera. In laboratory animal bacteriology the most important of these genera are *Pseudomonas* and *Bordetella*.

11.1 Motile, aerobic, Gram-negative rods

Five motile genera are of importance in laboratory animal bacteriology: *Pseudomonas, Agrobacterium, Burkholderia, Xanthomonas,* and *Chryseomonas*. In laboratory animal bacteriology *Bordetella* also belongs to this group, because the only important species, *B. bronchiseptica*, unlike human *Bordetella* spp., is motile. All genera can be differentiated according to Table 11.1. The commercial kit API 20NE (bioMérieux, France) may be used for identification and differentiation of all these species. This kit is based on assimilation tests, i.e., tests in which the ability of the isolate to use one single carbohydrate as energy source is tested. The kit is excellent for identification of *B. bronchiseptica* and *P. aeruginosa*, while identification of other *Pseudomonas* spp. should only be accepted with caution.

11.1.1 Pseudomonas

11.1.1.1 Characteristics of infection

Pseudomonas spp. are normal inhabitants of the environment and may contaminate tap water and even quarternary ammonium disinfectants. They may be isolated from the respiratory, digestive, and genital systems of rats and mice. As several of those species isolated are just passive inhabitants a variety of different *Pseudomonas* species may be observed, the more common ones being *P. aeruginosa* and *P. diminuta*. *P. aeruginosa* causes conjunctivitis and rhinitis and, under more severe or experimental conditions, pneumonia and septicemia in rats and guinea pigs. In septicemic animals abscesses of the liver, spleen, kidneys, and middle ear may be observed. Disease due to *P. aeruginosa* is mainly observed in immune-deficient, immunosuppressed, or stressed animals,[1] and, in general, it is secondary to something else. Ventilation breakdowns and other types of environmental stress may act as inducers. The prevalences in infected colonies of immune-competent animals kept in a high quality environment seldom reach more than 5 to 10%. The prevalence of diseased animals in colonies of immune-deficient animals kept

Table 11.1 Differentiation of Gram-Negative, Aerobic, Motile Rods Found in Rodents and Rabbits

	Pseudomonas spp.	*Bordetella bronchiseptica*	*Sphingomonas paucimobilis*	*Burkholderia cepacia*	*Xanthomonas maltophilia*	*Chryseomonas luteola*	*Agrobacterium radiobacter*
Colony pigments	Various pigments	White	Yellow	Nonpigmented on most media	Green to purple	Yellow	None
Oxidase	+	–	+	+	–	–	+
Esculin	–	–	+	d	+	+	+
Arginine dihydrolasis	+	–	–	–	–	–	–
Urease	d	+	–	d	–	d	+
Gelatine liquefaction	+	–	–	d	+	d	–
Maltose	–	–	+	+	+	+	+
Mannitol	d	–	–	+	–	+	+

under poor environmental conditions may reach 100%. Poor hygienic conditions, especially in relation to water used for drinking and cleaning, may play an important role in the spread of *Pseudomonas* spp. *P. fluorescens* and probably some other *Pseudomonas* spp. produce mucous in drinking nipples, which, however, is not known to have any impact on the animals. This condition is normally prevented by acidification of the drinking water with hydrochloric or citric acid.

11.1.1.2 Characteristics of the agent

11.1.1.2.1 Morphology. *Pseudomonas* are clearly Gram negative and appear as long, thin rods in young cultures and shorter rods in older cultures. Colonies usually are spreading on the agar surface and have a characteristic odor. As the culture gets older various types of pigmentation may occur. Cultivation of *P. aeruginosa* on King's agars (Table 11.2) at 35°C produces diffusable pigments.

Table 11.2 King's Agars A and B for Visualizing
Pigment Production by *Pseudomonas* spp.[a]

King's Agar A (Slants)

Sterile, deionized water	1000 ml
Peptone	20.0 g
Glycerol	8.7 g
Potassium sulfate	10.0 g
Magnesium sulfate, 6H$_2$O	3.0 g
Agar	15.0 g

Note: pH is stabilized at 7.2.

King's Agar B (Slants)

Sterile, deionized water	1000 ml
Peptone	20.0 g
Glycerol	8.7 g
Dipotassium phosphate	1.5 g
Magnesium sulfate, 7H$_2$O	1.5 g
Agar	15.0 g

Note: pH is stabilized at 7.4.

[a] If King's agar A turns blue the inoculated isolate is a pyocyanin-producing *P. aeruginosa*. King's agar B is inspected with a Woods lamp. If fluorescence is observed the inoculated isolate produces fluorescein, which is typical for *P. aeruginosa*, *P. fluorescens*, and *P. putida*.[12]

11.1.1.2.2 Cultivation.   A variety of media may be used. Nonselective, nonindicative media blood or chocolate agar is useful. MacConkey agar may be used to favor selection. Propagation media may be used, but they are usually not needed, as *Pseudomonas* spp. are easily isolated without propagation. Cultivation for 24 h at 37°C is sufficient. Cultivation at 42°C favors isolation of *P. aeruginosa*. Some species, e.g., *P. fluorescens,* may be favored by cultivation at temperatures as low as 4°C.

11.1.1.2.3 Isolation sites.   *Pseudomonas* spp. may be isolated from any sampling site on the animal, but the skin and sites within the respiratory system are preferable as primary sampling sites.

11.1.1.2.4 Differentiation and identification.   *Pseudomonas* spp. are easily differentiated from other aerobic, Gram-negative, motile rods according to Table 11.1. However, the differentiation between them may be more difficult. Simple biochemical tests as proposed in Table 11.3 may be applied. The commercial kits API 20NE and API 20E (bioMérieux, France) clearly identify *P. aeruginosa,* but are less reliable for other *Pseudomonas* spp.

11.1.1.2.5 Serology.   ELISA may be used to diagnose antibodies against *P. aeruginosa.*[2] In rats the sensitivity of cultivation and ELISA are similar.[3]

11.1.2 Xanthomonas, Sphingomonas, Chryseomonas, Agrobacterium, *and* Burkholderia

Xanthomonas maltophilia is occasionally isolated from rodents. It produces large pigmented colonies. A new name, *Stenotrophomonas maltophilia,* has been proposed.

Sphingomonas paucimobilis (formerly *Pseudomonas paucimobilis*) is occasionally isolated from the cecum of rodents. It is a long rod that forms large, yellow, butyrous, and sometimes mucoid colonies.

Chryseomonas luteola is occasionally isolated from the genitals and cecum of rodents, including guinea pigs and gerbils. It produces colonies that may be either wrinkled and adherent, or smooth and yellow.

Burkholderia cepacia (formerly *Pseudomonas cepacia*) is occasionally isolated from rabbits. It is a slow-growing organism that often needs 3 days to produce visible colonies.

Agrobacterium radiobacter is occasionally isolated from the respiratory system of rats. Colonies are 2 mm after 48 h of incubation. The colonies are non-pigmented.

The prevalences of these bacteria are low and no impact on the animals or research has been described. They may be identified on the basis of Table 11.1.

Table 11.3 Differentiation of *Pseudomonas* spp. Found in Rodents and Rabbits

	Pseudomonas								
	aeruginosa	*fluorescens*	*putida*	*stutzeri*	*mendocina*	*pseudo-alcaligenes*	*alcaligenes*	*diminuta*	*vesicularis*
Oxidase	+	+	+	+	+	+	+	+	+
Growth at 42°C	+	–	–	+	+	d	d	–	–
Nitrate reduction	+	d	–	+	+	d	d	–	–
Esculin	–	–	–	–	–	–	–	–	+
Lecithin	–	+	–	–	–	–	–	–	–
Arginine dihydrolasis	+	+	+	–	+	d	–	–	–
Gelatine liquefaction	+	d	–	–	–	–	–	d	d
Glucose	+	+	+	+	+	d	–	d	d
Lactose	–	d	d	–	–	–	–	–	–
Sucrose	–	d	d	–	–	–	–	–	–
Maltose	–	d	d	+	–	–	–	–	–
Mannitol	d	+	d	d	–	–	–	–	–

11.1.3 Bordetella

11.1.3.1 Characteristics of infection

Only one species, *B. bronchiseptica*, is of importance in laboratory animal bacteriology. It may be isolated from rabbits and guinea pigs and occasionally from rats, while mice, hamsters, and gerbils are less prone to infection. Disease, which is characterized by pneumonia, is most common in guinea pigs. Additionally, pleuritis and pericarditis may be observed. The disease is often fatal, but whether this is mostly so if the animal is also infected with other respiratory pathogens, such as parainfluenza virus type III, is not fully understood. In rabbits disease is mainly subclinical and characterized by focal chronic interstitial pneumonia.[4] In rats necrotizing bronchopneumonia may be observed.

11.1.3.2 Characteristics of the agent

11.1.3.2.1 Morphology.   The cells of *B. bronchiseptica* are coccobacilli or short rods 0.2 to 0.5 µm wide and 0.5 to 2.0 µm long. The colonies are small and convex with an entire margin, and may be β-hemolytic on blood agar.

11.1.3.2.2 Cultivation.   In contrast to human *Bordetella* spp., *B. bronchiseptica* easily grows on even simple media, i.e., blood agar, chocolate agar, and MacConkey agar. Occasionally, lactose-sucrose agars are used indicatively, as the nonfermenting small colonies of *B. bronchiseptica* can easily be differentiated from colonies produced by fermenting bacteria. Aerobic cultivation for 24 to 48 h at 37°C is sufficient.

11.1.3.2.3. Isolation sites.   The nose and trachea, and, in affected animals, the lungs, are suitable sampling sites.

11.1.3.2.4 Differentiation and identification.   *Bordetella* and *Brucella* are the only noncarbohydrate-utilizing, obligate aerobic, Gram-negative rods in laboratory animal bacteriology. *B. bronchiseptica* differs from *Brucella* spp. by its ability to grow on MacConkey agar, which *Brucella* does not. This is not a major problem in the identification process, as only *B. bronchiseptica* is a common finding in laboratory animal bacteriology. No carbohydrates are fermented. As no other members of *Bordetella* are to be expected in laboratory animals, differentiation between *B. bronchiseptica*, *B. pertussis*, *B. parapertussis*, and *B. avium* is not necessary. It is, however, easily done, as *B. bronchiseptica* is motile, which *B. pertussis* and *B. parapertussis* are not, and *B. bronchiseptica* reduces nitrate, which none of the others do. Furthermore, *B. bronchiseptica* is urease positive. The commercial kit API 20NE (bioMérieux, France) easily identifies *B. bronchiseptica*.

11.1.3.2.5 Serology.   Serology for *B. bronchiseptica* has traditionally been performed as an agglutination assay. A mixture of $2 \cdot 10^9$ bacteria per milliliter with 0.15% formalin and 1:10,000 thiomersal is used as antigen. The serum is diluted 1:20, 1:40, and so on. Serum of each dilution (0.2 ml) is added to 0.2 ml antigen in each of their test tubes and clear agglutination counts as positive.[5]

11.2 Nonmotile, aerobic, Gram-negative rods that can easily be cultivated

This group does not contain any bacteria known to have a serious impact on laboratory animals or research. The members may be identified according to Table 11.4. Alternatively, the commercial kit API 20NE (bioMérieux, France) may be used.

11.2.1 Acinetobacter

Acinetobacter may occur both as coccobacilli, coccoid rods, or more typical rods. Colonies are smooth, small, and nonpigmented. They are typically isolated from the respiratory system and genitals. *A. junii*, which grows at 41°C, is most often isolated from guinea pigs, while *A. lwoffii*, which never grows at 41°C, is quite common in rats.

11.2.2 Flavobacterium

Four species of *Flavobacterium* — *F. meningosepticum, F. odoratum, F. breve*, and *F. indologenes* — are occasionally isolated from laboratory rodents and rabbits. Although they may be part of a mixed flora in wounds, there is no indication that they should be of any importance as pathogens or have any severe impact on research. The cells are pleomorphic and *F. meningosepticum* and *F. indologenes* may show filamentous forms. Colonies are 1 to 2 mm and often pigmented.

11.2.3 Weeksella

Two species are of importance, *W. zoohelcum* and *W. virosa*, which occasionally may be isolated from the respiratory and genital systems of rats. They are not known to be of any importance. They grow easily on simple media. *W. virosa* grows at 42°C and on MacConkey agar, while *W. zoohelcum* neither grows at 42°C nor on MacConkey agar.

11.3 Nonmotile, aerobic, Gram-negative rods that are difficult to cultivate

In laboratory animal bacteriology, this group only contains one genus, *Francisella*, which, however, does not need to be considered in routine examinations.

Table 11.4 Differentiation of Nonmotile, Aerobic, Gram-Negative Rods Found in Laboratory Rodents and Rabbits

	Acinetobacter spp.	*Flavobacterium*				*Weeksella*	
		breve	*indologenes*	*meningosepticum*	*odoratum*	*zoohelcum*	*virosa*
Colony pigmentation	None	Pale yellow	Deep yellow	Very weak yellow or no pigmentation	Yellowish to tan	Tan to yellow	Tan to brown
Esculin	–	–	+	+	–	–	–
Urease	d	–	–	–	–	+	–
Indole	–	+	+	+	–	+	+
Maltose	d	+	+	+	+	–	–
Mannitol	–	–	–	+	–	–	–

11.3.1 Francisella

11.3.1.1 Characteristics of infection

Only one species, *F. tularensis*, is to be considered in this genus. This is the causative agent of tularemia in humans and wild animals. As it is found in wild rodents and rabbits, it is in principle also of some importance in laboratory animal bacteriology, although it is seldom diagnosed in laboratory rodents and rabbits. One case in a hamster colony actually has been described.[6] The first sign was a ruffled hair coat, but within 48 h the affected animals, which were 4 to 8 weeks old and weaned, were moribund. The outbreak was connected with feeding fresh vegetables, which should be banned in today's laboratory animal colonies.

11.3.1.2 Characteristics of the agent

Successful isolation has been achieved from ulcers, lymph nodes, gastric washing, and sputum, but isolation is extremely difficult. For primary isolation cystine glucose blood agar may be used.[7] Alternatively, chocolate agar supplemented with IsoVitale X (Becton Dickinson, France) and penicillin may be used. A heavy inoculum should be used which may lead to visible colonies within 4 days. The organisms are small and pleomorphic. Colonies are up to 2 mm, blue-gray, round, smooth, and slightly mucoid. On cystine glucose blood agar α-hemolysis is normally observed. Suspicious isolates are identified as *F. tularensis* by slide agglutination. Antibodies are commercially available (Difco, U.S.). Serology is probably the method of choice if routine examination is to be performed in laboratory animal colonies. Commercial antigen is available for tube agglutination test from either Difco (U.S.) or BBL (U.S.). Cutoff values of 1:80 or above should be used. Both ELISA[8] and polymerase chain reaction (PCR) on blood[9] have been described, but the sensitivity of the latter thus far seems to be rather low.

11.3.1.3 Safety

The U.S. CDC recommends Biosafety Level-2 practices, containment equipment, and facilities for activities with clinical materials of animal origin containing or potentially containing *F. tularensis*. Biosafety Level-3 and Animal Biosafety Level-3 practices, containment equipment, and facilities are recommended, respectively, for all manipulations of cultures and for experimental animal studies. An investigational live attenuated vaccine[10] is available. It is recommended for persons working with the agent or infected animals, and for persons working in or entering the laboratory or animal room where cultures or infected animals are maintained.[11]

References

1. Urano, T. and Maejima, K., Provocation of pseudomoniasis with cyclophosphamide in mice, *Lab. Anim.,* 12, 159, 1978.
2. Johansen, H. K., Espersen, F., Pedersen, S. S., Hougen, H. P., Rygaard, J., and Høiby, N., Chronic *Pseudomonas aeruginosa* lung infection in normal and athymic rats, *APMIS,* 101(3), 207, 1993.
3. Hansen, A. K., Improvement of health monitoring and the microbiological quality of laboratory rats, *Scand. J. Lab. Anim. Sci.,* 23(suppl. 2), 45, 1996.
4. Uzal, F. A., Feinstein, R. E., Rehbinder, C., and Persson, L., A study of lung lesions in asymptomatic rabbits naturally infected with *B. bronchiseptica, Scand. J. Lab. Anim. Sci.,* 16(1), 3, 1989.
5. Ohder, H. and Wullenweber, M., *Bordetella bronchiseptica,* in *Diagnostic Microbiology for Laboratory Animals,* Kunstyr, I., Ed., Gustav Fischer, New York, 1992, 57.
6. Permand, U. and Bergeland, M. E., A tularemia enzootic in a closed hamster breeding colony, *Lab. Anim. Care,* 17, 563, 1967.
7. Stewart, S.J., Tularemia, in *Diagnostic Procedures for Bacterial, Mycotic and Parasitic Infections* 6th ed., Balows, A. and Hausler, W.J., Jr., Eds., American Public Health Association, Washington, D.C., 1981, 705.
8. Yurov, S. V., Pchelintsev, S. Y., Afanasyev, S. S., Vorobyev, A. A., Urakov, N. N., Chenrenkova, G. V., Fedortsov, K. K., Krasnoproshina, L. I., Vlasov, G. S., and Denisova, N. B., The use of microdot immunoenzyme analysis with visual detection for the determination of tularemia antibodies, *Zh. Mikrobiol. Epidemiol. Immunobiol.,* 3, 64, 1991.
9. Long, G. W., Oprandy, J. J., Narayanan, R. B., Fortier, A. H., Porter, K. R., and Nacy, C. A., Detection of *Francisella tularensis* in blood by polymerase chain reaction, *J. Clin. Microbiol.,* 31, 152, 1993.
10. Burke, D.S., Immunization against tularemia: analysis of the effectiveness of live *Francisella tularensis* vaccine in prevention of laboratory-acquired tularemia, *J. Infect. Dis.,* 135, 55, 1977.
11. Office of Health and Safety, Centers for Disease Control and Prevention, Laboratory Biosafety Level Criteria, http://www.cdc.gov/od/ohs/biosfty/bmbl/section3.htm.
12. King, E.O., Ward, M. K., and Raney, D.E., Two simple media for the demonstration of pyocynanin and fluorescein, *J. Lab. Clin. Med.,* 44, 301, 1954.

chapter twelve

Obligate anaerobic
and microaerophilic
Gram-negative bacteria

Contents

12.1 Bacteroidaceae 204
    12.1.1 *Fusobacterium* 204
        12.1.1.1 Characteristics of infection...204
        12.1.1.2 Characteristics of the agent..204
            12.1.1.2.1 Morphology 204
            12.1.1.2.2 Cultivation...206
            12.1.1.2.3 Differentiation and identification...............206
12.2 Microaerophilic curved bacteria 206
    12.2.1 *Campylobacter* 206
        12.2.1.1 Characteristics of infection...206
        12.2.1.2 Characteristics of the agent..208
            12.2.1.2.1 Morphology..208
            12.2.1.2.2 Cultivation...208
            12.2.1.2.3 Differentiation and identification...............208
        12.2.1.3 Safety 209
    12.2.2 *Helicobacter* 209
        12.2.2.1 Characteristics of infection...209
        12.2.2.2 Characteristics of the agent..210
            12.2.2.2.1 Morphology 210
            12.2.2.2.2 Isolation sites and cultivation.....................210
            12.2.2.2.3 Differentiation and identification...............210
            12.2.2.2.4 Serology...211
            12.2.2.2.5 Molecular biology...211
12.3 Cilia-associated respiratory bacillus...211
    12.3.1 Characteristics of infection...211

            12.3.2 Characteristics of the agent...212
                    12.3.2.1 Isolation...212
                    12.3.2.2 Identification..212
References...212

12.1. Bacteroidaceae

Anaerobic Gram-negative bacteria are found in the cecum of laboratory animals and may be given to the animals as a part of the defined flora used for gnotobiotic animals after rederivation. This is a rather unexplored field, and, therefore, it is not always possible to characterize fully isolates discovered by anaerobic cultivation from laboratory animals. The genera, *Bacteroides* and *Fusobacterium*, are both known to be found in rodents and rabbits, but to which extent other members of this group are also found is not fully clarified. One species, *F. necrophorum*, is pathogenic and causes necrosis in rabbits.

The simplest way to divide Gram-negative anaerobic rods is to test them for their antibiotic susceptibility, indole, urease, alkaline phosphatase, and xylose fermentation (Table 12.1). It should be noted that species other than those listed in Table 12.1 are defined in systematic bacteriology, and there is no knowledge whether these may be found in rodents or rabbits, too. Several commercial kits, e.g., AniCard, API 20A, and RAPID ID 32A from bioMérieux (France) or Minitek from BBL (U.S.) are helpful to some extent. The antibiotic and enzymatic tests proposed in Table 12.1 may be purchased commercially as well, from Rosco (Denmark), for example.

12.1.1 Fusobacterium

12.1.1.1 Characteristics of infection

In this context *Fusobacterium* is of importance only in the rabbit. *F. necrophorum* is not the only species within this genus, but it is the main pathogen. In rabbits *F. necrophorum* may produce either a phlemognous and purulent necrosis, typically starting in the nasal region, or a chronic suppuration over the legs and flank. Such clinical symptoms are not very common in laboratory rabbits today, but as routine screenings for this agent are seldom performed in rabbit colonies, little is known about whether laboratory rabbits actually still harbor this organism. Whether laboratory animals also harbor some of the other species in the genus is not known.

12.1.1.2 Characteristics of the agent

12.1.1.2.1 Morphology.   Bacterial morphology is pleomorphic. Coccoid as well as longer filamentous rods may be observed. Metachromatic granules

Table 12.1 Differentiation of Anaerobic Gram-Negative Rods

	Kanamycin (1000 µg)	Vancomycin (5 µg)	Colistin (10 µg)	Growth in 20% bile	Indole	Alkaline phosphatase	Urease	Xylose
Bacteroides fragilis group	R	R	R	+	d	+	–	+
B. ureolyticus group	S	R	S	–	–	d	+	–
B. gracilis[a]	S	R	S	–	–	+	–	?
Bacteroides (other spp.)	R	R	d	–	+	+	–	+
Porphyromonas asaccharolytica	R	S	R	–	+	+	–	–
P. gingivalis	R	S	R	–	+	+	–	+
Prevotella intermedia	R	R	S	–	+	+	–	–
Prevotella (other spp.)	R	R	d	–	–	+	–	–
Fusobacterium necrophorum	S	R	S	–	+	+	–	–
F. nucleatum	S	R	S	–	+	–	–	–
F. varium	S	R	S	+	+	–	–	–
F. mortiferum	S	R	S	+	–	+	–	–

[a] *B. gracilis* is not a fully defined species and may be separated into one or several new genera in the future.

may be observed. Colonies are white, small, and round. If human or rabbit blood is used colonies on blood agar are hemolytic.

12.1.1.2.2 Cultivation.   Chocolate agar with the addition of vitamin K and cysteine (Table 3.2) should be used for primary isolation. The inoculated medium is incubated anaerobically at 37°C for 24 to 48 h. Pure cultures may be grown on blood agar containing rabbit or human blood. Sampling is performed from the deep parts of the necrotic area in affected rabbits. From healthy animals scrapings of the inner lining of the cecal mucosa should be used.

12.1.1.2.3 Differentiation and identification.   Identification is carried out according to Table 12.1. *Fusobacterium* spp. is characterized by resistance to vancomycin, but sensitivity to both kanamycin and colistin. *F. necrophorum* may be further divided into biovars: biovar A/F, also called *F. necrophorum* subsp. *necrophorum*, produces hemagglutinin and lipase, which biovar B, also called *F. necrophorum* subsp. *funduliforme*, does not.

12.2 Microaerophilic curved bacteria

This group in up-to-date systematics contains two genera, *Campylobacter* and *Helicobacter*; these are bacteria that are found primarily in the digestive system and some related organs, such as the liver. Previously both belonged to the same genus, *Campylobacter*, but in 1989 the genus *Helicobacter* was established by reclassification of the genus *Campylobacter* on the basis of DNA analysis.[1] Differentiation between the two may be somewhat difficult in the routine laboratory. In human bacteriology, the urease test is the most applicable, as human *Helicobacter* spp. are urease positive, while all *Campylobacter* spp. are urease negative. However, rodent *Helicobacter* species, which are urease negative, do exist and therefore the urease test has to be combined with hippurate hydrolysis, growth at 42°C, and antibiotic susceptibility testing (Table 12.2).

12.2.1 Campylobacter

12.2.1.1 Characteristics of infection
C. coli and *C. jejuni* have been isolated from the digestive system of mice, rats, hamsters, and rabbits. Humans and some farm animal species are known to develop severe disease symptoms in relation to infection with *Campylobacter*, but this is not known to be the case in rodents and rabbits, nor is it clear whether *Campylobacter* infections in these species have a zoonotic potential. The prevalence rates in infected colonies are rather low, typically less than 5%.

Table 12.2 Differentiation of *Helicobacter* and *Campylobacter* spp. from Rodents

	Helicobacter							Campylobacter	
	cinaedi	*bilis*	*hepaticus*	*muridarum*	*rodentium*	*rappini*	*trogontum*	*jejuni*	*coli*
Nitrate reduction	+	+	+	–	+	–	+	+	+
Alkaline phosphatase	–	?	?	+	–	–	–	d	d
Urease	–	+	+	+	–	+	+	–	–
Hippurate hydrolysis	–	–	–	–	–	–	–	+	–
γ-Glutamyl transferase	–	?	?	+	–	+	+	d	–
Growth at 42°C	–	+	–	–	+	+	+	+	+
Nalidixic acid (30 µg)	S	R	R	R	R	R	R	S	S
Cephalothin (30 µg)	I	R	R	R	R	R	R	R	R

Some data from Fox, J. G. and Lee, A., *Lab. Anim. Sci.*, 47(3), 222, 1997.

12.2.1.2 Characteristics of the agent

12.2.1.2.1 Morphology. *Campylobacter* are arching, motile rods, often appearing as a comma, or, if grouped in pairs, as a sea gull (Figure 4.1[10]). In older cultures the cells may become coccoid. Colonies on TVP agar (Table 12.3) are small, either gray or transparent, and nonhemolytic. They may also appear as a thin film on the agar, however, in a thicker layer than the films created by *Helicobacter* spp.

Table 12.3 TVP Agar for the Isolation of *Campylobacter* and *Helicobacter* spp.

Sterile, deionized water	1000 ml
Magnesium sulfate, $7H_2O$	0.1 g
Mangane chloride, $4H_2O$	0.007 g
Disodium hydrogen phosphate, $12H_2O$	8.0 g
Caseine hydrolysate	5.0 g
Yeast extract	3.0 g
Peptone	5.0 g
Potassium chloride	6.7 g
Sebacic acid	0.01 g
Agar	10.0 g
Cystein HCl	0.05 g
Defibrinized horse blood	50 ml
Sodium pyruvate	2.0 g
Trimethoprim	0.005 g
Vancomycin	0.01 g
Polymyxin B	2500 i.u.

Note: pH is stabilized at 7.2.

Røder, B. L., *Substrathåndbogen* [Handbook of Substrates], Statens Serum-institut, Copenhagen, 1993, 24. With permission.

12.2.1.2.2. Cultivation. Feces and stool may be sampled from live animals. From euthanized animals cecal scrapings, gallbladder (do note that the rat does not have a gallbladder!), as well as intestinal contents may be cultivated as described for *Helicobacter* spp. (see below) using a 0.65-µm filter. TVP agar (Table 12.3) may be used. Propagation in an enrichment broth has also been described,[2] and the broth is commercially available (Oxoid, U.K.).

Incubation must be microaerophilic for at least 48 h but preferably for 3 to 5 days at 37°C, either sealed with a microaerophilic gas kit or in a CO_2 incubator (see Chapter 3). The optimal atmosphere is 10% CO_2, 85% N_2, and 5% O_2.

12.2.1.2.3 Differentiation and identification. Identification is carried out according to Table 12.2. Alternatively, the commercial kit API Campy (bioMérieux, France) is applicable. Two immunological kits, Campyslide

(BBL, U.S.) and Meritec Campy Jcl (Meridian Diagnostics, U.S.) identify *C. jejuni* and *C. coli*, but without differentiating between the two. So far, neither polymerase chain reaction nor serology seems to be applicable in laboratory animal health monitoring for the detection of *Campylobacter* spp.

12.2.1.3 Safety

The U.S. CDC recommends Biosafety Level-2 practices, containment equipment, and facilities for activities with cultures of, or clinical materials potentially infected with, *C. coli* and *C. jejuni*. Animal Biosafety Level-2 practices, containment equipment, and facilities are recommended for activities with naturally or experimentally infected animals. Vaccines are not available for use in humans.[3]

12.2.2 Helicobacter

12.2.2.1 Characteristics of infection

Several rodent *Helicobacter* species have been isolated (Table 12.4). Clinical significance has only been documented in relation to a few of these.

H. hepaticus is known to cause chronic hepatitis and probably also liver tumors in mice. Susceptibility to disease seems to be genetically dependent, e.g., A/JCr mice seem to be highly susceptible, while C57BL mice seem to be resistant.[4]

H. bilis probably also causes hepatitis, although this is not as clearly documented as for *H. hepaticus*. Also, *H. cholecystus* may cause hepatitis and pancreatitis in hamsters.

The competition between spontaneous and experimental *Helicobacter* infection, when rodents are used as models for human infection with *H. pylori*, may be an important factor if experimental infection fails.

Table 12.4 *Helicobacter* spp. in Rodents and Their Clinical Importance

Name	Species	First report	Clinical significance
H. cinaedi	Hamsters	1989[28]	Not reported
H. muridarum	Rats, mice	1992[29]	Not reported
H. rappini	Mice	1993[30]	Not reported
H. hepaticus	Mice	1994[31]	Chronic hepatitis, probably liver tumors
H. bilis	Mice	1995[32]	Probably hepatitis
H. trogontum	Rats	1996[33]	Not reported
H. cholecystus	Hamsters	1996[34]	Probably hepatitis and pancreatitis
H. rodentium	Mice	1997[35]	Not reported

At present, it is not clear how widespread infection with *Helicobacter* spp. is in rodent colonies, as routine investigations have only been performed at a few laboratories. However, shortly after the first reports on *Helicobacter* infections, the agent was isolated from some major breeding and reference centers for mice, and, therefore, it is very likely that a great number of mouse colonies may be infected with this organism. It is also likely that new rodent *Helicobacter* spp. may be identified if anaerobic cultivation regimes stricter than those standard in rodent health monitoring laboratories today could be applied, as it is known from other species that some *Helicobacter* spp. do not tolerate oxygen at all, not even during the handling of cultures between incubations.

12.2.2.2 Characteristics of the agent

12.2.2.2.1 Morphology. *Helicobacter* spp. are helical, curved, or straight Gram-negative rods, which are approximately 0.3 to 1 µm wide and 1 to 5 µm long (Figure 4.1[6]). Bacterial growth upon an agar plate appears as a thin film, which may not be observed if the plates are not properly inspected.

12.2.2.2.2 Isolation sites and cultivation. The most optimal sampling sites in mice are the liver and the inner lining of the cecum. From live mice fecal pellets may be used. Sampling for *H. muridarum* in mice should also involve the stomach mucosa. From the liver a direct smear on the agar plate may be effective, while the cecum, stomach, and fecal pellets have to be sampled with great care. The inner lining is scraped with a sterile scalpel and the scraping is then immersed in sterile 0.9% saline, which is whirl-mixed. The immersion is then filtered directly upon an agar plate or into an enrichment broth. The pore size of the filter should be 0.45 µm for *H. hepaticus*, but 0.65 µm for *H. bilis*, *H. trogontum*, and *H. rappini*.

Chocolate agar seems to support growth. From cecum it may be desirable to use a selective medium, such as TVP agar (Table 12.3), although it does not support growth as much as chocolate agar does. The medium is incubated for 3 to 5 days at 37°C, either sealed with a microaerophilic gas kit or in a CO_2 incubator (see Chapter 3). The optimal atmosphere is 5% CO_2, 90% N_2, and 5% H_2.

12.2.2.2.3 Differentiation and identification. Broths to be used for tests for biochemical properties in *Helicobacter* spp. must be supplemented with 10% fetal calf serum. Growth is facilitated if the flask is placed in a Gyrotory water bath shaker (New Brunswick Scientific Co., U.S.) at 150 rpm, fitted with a gassing hood connected to a microaerophilic gas mixture.[5] The commercial kit API Campy (bioMérieux, France) may be helpful in the identification process, although the computerized identification procedure does not reveal rodent *Helicobacter* spp. Results from in-house tests as well as the commercial kit may be interpreted using Table 12.2.

12.2.2.2.4 Serology. ELISA may be used to detect antibodies against *H. hepaticus*, and antigens are commercially available from Harlan (U.K.). However, the sensitivity may be low and correlated to the degree of hepatitis. Males above the age of 12 months or females above the age of 18 months should be sampled. The assay only detects antibodies to *H. hepaticus* and not to other rodent *Helicobacter* spp., unless these are used as antigens.

12.2.2.2.5 Molecular biology. Polymerase chain reaction (PCR) is currently the most rapid and perhaps the most sensitive method of diagnosis for *H. hepaticus*. PCR may also be used to identify *H. hepaticus, H. bilis, H. muridarum,* and *H. rappini* in the same amplification procedure, after which identification to a species level may be done by restriction enzyme analysis.[6] PCR for *Helicobacter* spp. is conducted on fecal pellets, cecal contents, or liver homogenates.[7] A purification procedure is necessary to prevent inhibition.[8]

12.3 Cilia-associated respiratory bacillus

12.3.1 Characteristics of infection

Infection with cilia-associated respiratory (CAR) bacillus has been reported in mice, rats, and rabbits, and in farm animals such as pigs and goats, in Europe, the U.S., and Japan.

CAR bacillus isolates from — on the one hand — rats and mice and — on the other hand — rabbits are host specific and should be regarded as different bacteria that belong to distinct genera. Determination of gene sequences have shown that CAR bacillus from rats and from mice are related to *Flavobacterium*,[9] while the rabbit isolates show a greater similarity with *Helicobacter*.[10]

The pathogenicity is not fully clarified and infected rodents are usually asymptomatic,[11,12] but this agent is claimed to be the cause of *chronic respiratory disease* in rats, i.e., a highly contagious epizootic, slowly progressive, and uncontrollable disease, in which symptomatic rats histopathologically show various degrees of pulmonary changes, such as mucopurulent exudate and severe peribronchial lymphoid cuffing.[13] Clinical signs, if present, may include weight loss, rough hair coat, wheezing, and rales. The same clinical picture may be observed in mice,[14] although the possible involvement of other pathogens should also be considered. There are no differences in the susceptibility between rat strains, but different strains of the agent are more or less virulent. Severity of symptoms during experimental infection seems to be worse if an isolate of the same species is used.[11,15] In rabbits no clinical signs of respiratory disease have been observed, although histopathologic examination of the respiratory tree may reveal mild hyperplasia of lymphoid nodules subjacent to the respiratory mucosa with scattered CAR bacilli in the lower respiratory tract.[16] Contact infection may play a major role in the transmission of this disease.

12.3.2 Characteristics of the agent

12.3.2.1 Isolation

Cultivation in traditional bacteriological media is difficult. For several years propagation by inoculation of embryonated chicken eggs via the allantoic route has been used. The agent may also be propagated in Dulbecco's or Eagle's minimum essential medium supplemented with 10% fetal calf serum. Additionally, 20% hamster tracheal organ culture soup may be added. Care should be taken to avoid contamination of the isolates with *Mycoplasma*.[17]

Isolates from rats are approximately 0.2 µm wide and 4 to 8 µm long with a triple-layer cell wall and bulbous ends. Cells are Gram negative, nonacid fast, nonspore forming, and heat labile (56°C for 30 min). It is motile but lacks structures resembling flagella, pili, or axial filaments. The rabbit bacilli are smaller and form fewer aggregates during propagation. The propagated bacteria may be stored at –70°C.[18–20]

12.3.2.2 Identification

In paraffin-embedded infected lungs the filamentous bacteria may be detected on the border of the tracheal and bronchial epithelium by various silver stains, e.g., Warthin-Starry, or more specifically by using either indirect immunofluorescence or immunoperoxidase staining technique based on the labeled streptavidin biotin method and 3-amino-9-ethylcarbazole (AEC) as substrate[21] (see Chapter 5).

Antibodies to CAR bacillus may be detected in the early stage of the infection. The first method developed was IFA using tracheal sections of infected mice as antigen.[22] The most common serological assay today is ELISA. The antigen is commercially available from Harlan (U.K.). It may also be prepared on site.[23] Allantoic fluids containing approximately $2 \cdot 10^8$ per milliliter of the agent are washed with sterile phosphate buffered saline (PBS), resuspended 1:5 in sterile coating buffer, and sonicated. Dilutions (1:40 or 1:80) are used as antigen solutions and coated on the surface of microtiter plates as described in Chapter 5.

CAR bacillus may be detected by PCR on infected lung tissue[24] or on swabs from the oral cavity already on day 3 postexposure.[25]

References

1. Goodwin, C. S., Armstrong, J. A., Chilvers, T., Peters, M., Collins, M. D., Sly, L., McConell, W., and Harper, W. E. S., Transfer of *Campylobacter pylori* and *Campylobacter mustelae* to *Helicobacter* gen. nov. as *Helicobacter pylori* comb. nov. and *Helicobacter mustelae* comb. nov., *Int. J. Syst. Bacteriol.*, 39, 397, 1989.
2. Martin, W. T., Patton, C. M., Morris, G. K., Potter, M. E., and Puhr, N. D., Selective enrichment broth for the isolation of *Campylobacter jejuni*, *J. Clin. Microbiol.*, 17, 853, 1983.

3. Office of Health and Safety, Centers for Disease Control and Prevention, Laboratory Biosafety Level Criteria, http://www.cdc.gov/od/ohs/biosfty/bmbl/section3.htm.
4. Ward, J. M., Anver, M. R., Haines, D. C., and Benveniste, R. E., Chronic active hepatitis in mice caused by *Helicobacter hepaticus*, *Am. J. Pathol.*, 145, 959, 1994.
5. Morgan, D. R., Freedman, C. E., Depew, C. E., and Kraft, W. G., Growth of *Campylobacter pylori* in liquid media, *J. Clin. Microbiol.*, 25, 2123, 1987.
6. Riley, L. K., Franklin, C. L., Hook, R. R., and Besch-Williford, C., Identification of murine helicobacters by PCR and restriction enzyme analysis, *J. Clin. Microbiol.*, 34, 942, 1996.
7. Charles River Laboratories Inc., Helicobacter Infection in Laboratory Mice: History, Significance, Detection and Management, http://www.criver.com/tech-docs/helico-1.html.
8. Shames, B. J., Fox, J. G., Dewhirst, F. E.,Yan, L., Shen, Z., and Taylor, N. S., Identification of widespread *H. hepaticus* infection in feces in commercial mouse colonies by culture and PCR assay, *J. Clin. Microbiol.*, 33, 2968, 1995.
9. Wei, Q., Tsuji, M., Takahashi, T., Ishihara, C., and Itoh, T., Taxonomic status of car bacillus based on the small subunit ribosomal RNA sequences, *Chin. Med. Sci. J.*, 10(4), 195, 1995.
10. Cundiff, D. D., Besch-Williford, C. L., Hook, R. R., Jr., Franklin, C. L., and Riley, L. K., Characterization of cilia-associated respiratory bacillus in rabbits and analysis of the 16S rRNA gene sequence, *Lab. Anim. Sci.*, 45(1), 22, 1995.
11. Shoji, Y., Itoh, T., and Kagiyama, N., Pathogenesis of two CAR bacillus strains in mice and rats, *Exp. Anim.*, 37, 447, 1988.
12. Cundiff, D. C. and Besch-Williford, C., Respiratory disease in a colony of rats, *Lab. Anim.*, 21, 16, 1992.
13. Itoh, T., Kohyama, K., Takakura, A., Takenouchi, T., and Kagiyama, N., Naturally occurring CAR bacillus infection in a laboratory rat colony and epizootiological observations, *Jikken Dobutsu*, 36(4), 387, 1987.
14. Griffith, J. W., White, W. J., Danneman, P. J., and Lang, C. M., Cilia-associated respiratory (CAR) bacillus infection of obese mice, *Vet. Pathol.*, 25(1), 72, 1988.
15. Schoeb, T. R., Davidson, M. K., and Davis, J. K., Pathogenicity of cilia-associated respiratory (CAR) bacillus isolates for F344, LEW, and SD rats, *Vet. Pathol.*, 34(4), 263, 1997.
16. Waggie, K. S., Spencer, T. M., and Allen, A. M., Cilia-associated respiratory bacillus infection in New Zealand rabbits, *Lab. Anim. Sci.*, 37, 533, 1987.
17. Schoeb, T. R., Dybvig, K., Keisling, K. F., Davidson, M. K., and Davis, J. K., Detection of *Mycoplasma pulmonis* in cilia-associated respiratory bacillus isolates and in respiratory tracts of rats by nested PCR, *J. Clin. Microbiol.*, 35(7), 1667, 1997.
18. Schoeb, T. R., Dybvig, K., Davidson, M. K., and Davis, J. K., Cultivation of cilia-associated respiratory bacillus in artificial medium and determination of the 16S rRNA gene sequence, *J. Clin. Microbiol.*, 31(10), 2751, 1993.
19. Shoji, Y., Itoh, T., and Kagiyama, N., Propagation of CAR bacillus in artificial media, *Jikken Dobutsu*, 41(2), 231, 1992 .
20. Ganaway, J. R., Spencer, T. H., Moore, T. D., and Allen, A. M., Isolation, propagation, and characterization of a newly recognized pathogen, cilia-associated respiratory bacillus of rats, an etiologic agent of chronic respiratory disease, *Infect. Immun.*, 47, 472, 1985.

21. Oros, J., Matsushita, S., Rodriguez, J. L., Rodriguez, F., and Fernandez, A., Demonstration of rat CAR bacillus using a labelled streptavidin biotin (LSAB) method, *J. Vet. Med. Sci.*, 58(12), 1219, 1996.

22. Matsushita, S., Kashima, M., and Joshima, H., Serodiagnosis of cilia-associated respiratory bacillus infection by the indirect immunofluorescence assay technique, *Lab. Anim.*, 21(4), 356, 1987.

23. Shoji, Y., Itoh, T., and Kagiyama, N., Enzyme-linked immunosorbent assay for detection of serum antibody to CAR bacillus, *Jikken Dobutsu*, 37(1), 67, 1988.

24. Cundiff, D. D., Besch-Williford, C., Hook, R. R., Jr., Franklin, C. L., and Riley, L. K., Detection of cilia-associated respiratory bacillus by PCR, *J. Clin. Microbiol.*, 32(8), 1930, 1994.

25. Goto, K., Nozu, R., Takakura, A., Matsushita, S., and Itoh, T., Detection of cilia-associated respiratory bacillus in experimentally and naturally infected mice and rats by the polymerase chain reaction, *Exp. Anim.*, 44(4), 333, 1995.

26. Fox, J. G. and Lee, A., The role of *Helicobacter* species in newly recognized gastrointestinal diseases of animals, *Lab. Anim. Sci.*, 47(3), 222, 1997.

27. Skirrow, M. B., *Campylobacter* enteritis: a "new" disease, *Br. Med. J.*, 2, 9, 1977.

28. Stills, H. F., Hook, R.R., and Kinden, D. A., Isolation of *Campylobacter*-like organism from healthy Syrian hamsters (*Mesocricetus auratus*), *J. Clin. Microbiol.*, 27, 2497, 1989.

29. Lee, A., Phillips, M. W., O'Rourke, J. L., Paster, B. J., Dewhirst, F. E., Fraser, G. J., Fox, J. G., Sly, L. I., Romaniuk, L. I., Trust, T. J., and Kouprach, S., *Helicobacter muridarum* sp. nov., a microaerophilic helical bacterium with a novel ultrastructure isolated from the intestinal mucosa of rodents, *Int. J. Syst. Bacteriol.*, 42, 27, 1992.

30. Schauer, D. B., Ghori, N., and Falkow, S., Isolation and characterization of "*Flexispira rappini*" from laboratory mice, *J. Clin. Microbiol.*, 31, 2709, 1993.

31. Fox, J. G., Dewhirst, F. E., Tully, J. G., Paster, B. J., Yan, L., Taylor, N. S., Collins M. J., Jr., Gorelick, P. L., and Ward, J. M., *Helicobacter hepaticus* sp. nov., a microaerophilic bacterium isolated from livers and intestinal mucosal scrapings from mice, *J. Clin. Microbiol.*, 32, 1238, 1994.

32. Fox, J. G., Yan, L. L., Dewhirst, F. E., Paster, B. J., Shames, B., Murphy, J. C., Hayward, A., Belcher, J. C., and Mendes, E. N., *Helicobacter bilis* sp. nov., a novel *Helicobacter* species isolated from the bile, liver, and intestines of aged, inbred mice, *J. Clin. Microbiol.*, 33, 445, 1995.

33. Mendes, E.N., Queiroz, D. M. M., Dewhirst, F. E., Paster, B. J., Moura, S. B., and Fox, J. G., *Helicobacter trogontum* sp. nov., isolated from the rat intestine, *Int. J. Syst. Bacteriol.*, 46, 916, 1996.

34. Franklin, C. L., Beckwith, C. S., Livingstone, R. S., Riley, L. K., Gibson, S.V., Besch-Williford, C. L., and Hook, R. R., Jr., Isolation of a novel *Helicobacter* species, *Helicobacter cholecystus* sp. nov., from the gall bladder of Syrian hamsters with cholioangiofibrosis and centrilobular pancreatitis, *J. Clin. Microbiol.*, 34, 2952, 1996.

35. Shen, Z., Fox, J. G., Dewhirst, F. E., Paster, B. J., Foltz, C. J., Yan, L., Shames, B., and Perry, L., *Helicobacter rodentium* sp.nov., a urease negative *Helicobacter* spp. isolated from laboratory mice, *Int. J. Syst. Bacteriol.*, 47(3), 627, 1997.

chapter thirteen

Spirochetes

Contents

13.1 *Treponema* 216
    13.1.1 Characteristics of infection 216
    13.1.2 Characteristics of the agent 216
        13.1.2.1 Morphology 216
        13.1.2.2 Microscopy 216
        13.1.2.3 Cultivation 217
        13.1.2.4 Serology 217
        13.1.2.5 Molecular biology 217
13.2 *Leptospira* 218
    13.2.1 Characteristics of infection 218
    13.2.2 Characteristics of the agent 218
        13.2.2.1 Microscopy 218
        13.2.2.2 Cultivation 219
        13.2.2.3 Differentiation and identification 219
        13.2.2.4 Serology 220
    13.2.3 Safety 220
13.3 *Spirillum minus* 220
References 220

Spirochetes are motile, spiral bacteria, which are difficult to cultivate in ordinary media. They may be divided into two families:

- *Spirochaetaceae*, which contains the four genera *Spirochaeta*, *Cristispira*, *Borrelia*, and *Treponema*, of which only the latter is of importance in laboratory animal bacteriology
- *Leptospiraceae*, which contains one genus, *Leptospira*, which is, however, important in rodent bacteriology

A third organism, *Spirillum minus*, is briefly mentioned in this chapter. Because it cannot be isolated and cultivated, however, classification is difficult.

13.1 Treponema

13.1.1 Characteristics of infection

Fourteen species of *Treponema* have been described. None of these are of any importance in rodents, while one species, *T. cuniculi*, the causative organism of rabbit syphilis, must be considered in rabbits. However, although infection with *T. cuniculi* is nearly exclusively limited to domestic and wild rabbits, latent infections may in rare cases be found in mice, guinea pigs, and hamsters. Clinical syphilis in rabbits is a painful condition primarily localized to/around the genitals. Inflammation starts with development of edema in the vulva, vagina, or prepuce, which over some time develops into vesicles, which may turn into ulcers. In severe and more chronic cases, erosions may appear in the head region, as well. In contrast to human syphilis caused by *T. pallidum* rabbits do not suffer from generalized cases involving the central nervous system, and, in fact, most cases are latent with no clinical symptoms. Spread is primarily venereal, and, therefore, the prevalence is highest in breeding colonies. Infection with *T. cuniculi* may disturb the use of rabbits in human syphilis research and diagnostic work. Today, most colonies of laboratory rabbits bred under acceptable hygienic conditions should be free of the organism.

Although *T. pallidum* and *T. cuniculi* cannot be differentiated either morphologically or serologically, *T. cuniculi* is not considered of any zoonotic importance.

13.1.2 Characteristics of the agent

13.1.2.1 Morphology

T. cuniculi is approximately 0.18 μm wide and 6 to 15 μm long. It is regularly helical (Figure 4.1[6]). It cannot be seen in ordinary light microscopy without staining.

13.1.2.2 Microscopy

Dark-field microscopy is the method of choice for observing motility in fresh preparations, but phase-contrast microscopy may also be used. Rapid rotation about the axis may be observed. The agent may also be observed in Giemsa, India ink (contrast), or immunofluorescense (IF) stains. The latter is obviously helpful as a diagnostic tool, but of course none of the three methods are able to show motility.

To prepare a slide for microscopy the crustae of an ulcer should be removed and the wound carefully cleaned with physiological saline. After drying the wound with gauze, some wound fluid is pressed out. A drop is mixed with physiological saline or tap water on a slide.

Slides to be used for Giemsa or IF stains are air dried. Fixation and IF-staining are then performed as described in Chapter 5. Slides for Giemsa staining are placed in methanol for 10 s and then in Giemsa solution (Merck, Germany) for 18 h. The staining solutions should be changed several times.[1]

For dark-field microscopy a coverslip is placed upon the immersion and the preparation is microscoped within 20 min. First, observations are done through the 400× magnification objective. When an organism has been localized a drop of immersion oil is placed upon the cover slip and motility is observed at 1000× magnification.

13.1.2.3 Cultivation

T. cuniculi cannot be cultured in ordinary bacteriological media, probably due to rather specific demands on the atmosphere. Inoculation of specimens into noninfected rabbits has been used continuously for propagation of *T. cuniculi*. It has been shown that *T. pallidum* is microaerophilic rather than anaerobic and that it can be propagated at least for some successive generations in complex cell media if the atmosphere contains only 1 to 5% O_2 and the medium contains reducing agents.[2] Such cultivation does not seem to be of much use in routine diagnostic work.

13.1.2.4 Serology

Antibodies are easily diagnosed in sera from infected rabbits. As there are no serological differences commercial kits for *T. pallidum* are fully usable for rabbits.[3] The method of choice is the immunofluorescence assay (IFA) (see Chapter 5). Antigen is commercially available, e.g., from Sanofi Pasteur (U.S.) or Lee Laboratories (U.S.). A less sensitive but occasionally more specific alternative is hemagglutination. Erythrocytes sensitized with antigen are agglutinated in a microtiter plate with U-wells. Dilution series are made if the test is to be quantitative. The value of using non-treponemal tests — such as the cardiolipin test used for human patients — for screening healthy rabbits is not fully clarified. ELISA for human diagnosis has been described,[4] but these are based on IgM screening, i.e., they mostly detect acute infections and are as such of no use in routine screening of healthy animals.

13.1.2.5 Molecular biology

Polymerase chain reaction (PCR) is usable for direct detection of *T. pallidum* in human patients.[5] There is no reason to believe that such PCR should not also be applicable to diagnose syphilis in rabbits. However, as it is a rather simple task to sample directly from the ulcers followed by, e.g., IF-staining, the ability to diagnose latent infection by PCR would be more useful.

13.2 Leptospira

13.2.1 Characteristics of infection

The genus *Leptospira* consists of 11, perhaps 12, species. Leptospirosis is a zoonotic disease in humans and dogs caused only by *L. interrogans*, of which at least 180 serotypes have been described, which again are grouped in 23 serogroups. In humans disease is characterized by high fever, occasionally by subserosal hemorrhage, and seldom by jaundice. The serovar *ballum* is found in mice, while the serovar *icterohemorrhagiae* is found in rats. These are latent infections, which are mostly found in wild animals that carry the agent in their renal tubules, from where it is shed in the urine. Humans are infected after contact with such animals or their urine, or — more commonly — after contact with contaminated water or soil. The agents enter the patient through breaks in the skin or the mucosal or conjunctival surfaces.

Laboratory rodents from properly protected and health-monitored colonies should be free of this agent.

13.2.2 Characteristics of the agent

13.2.2.1 Microscopy

Leptospira spp. are observed at best in dark-field or phase-contrast microscopy. The urine is the most obvious place to search for these agents in rodents. Before microscopy the sample should be centrifuged at $1500 \cdot g$ for 30 min. The supernatant is discharged and a wet mount is prepared from the sediment. Alternatively, the kidneys may be crushed in a mortar and suspended 1:10 in PBS. The suspension is centrifuged at $500 \cdot g$ for 15 min and the sediment is discharged, while the supernatant is recentrifuged at $1500 \cdot g$ for 30 min and prepared as for urine samples.

In dark-field or phase-contrast microscopy these motile organisms are helical rods, approximately 0.1 µm wide and 6 to 12 µm long (Figure 4.1[6]). The ends are hooked and swellings are often observed. The movements are rotating and slowly gliding.

For diagnosis immunofluorescence staining may be applied. The method is performed as described in Chapter 5, but the slides prepared with the sediment sampled as described above should be fixed with absolute alcohol for 10 min, and a fluorescein-isothiocyanate-conjugated antileptospiral serum may be applied in a direct assay. Such antibodies are available from either Difco (U.S.) or the Animal Plant and Health Inspection Services (U.S.). The diluted antibody should be placed on the slide for 2 h at 37°C.[6]

As a simple method Giemsa staining may be applied. Undiluted Giemsa solution (1 ml) (Merck, Germany) is added to 50 ml distilled water. Slides as fixed for immunofluorescence staining are stained in the solution at 37°C for 8 h.[7]

DNA amplification methods are under development, but so far the methods are no more sensitive than, e.g., immunofluorescence staining.

13.2.2.2 Cultivation

One or two drops of urine or an aseptically prepared kidney suspension may be inoculated into a semisolid medium, such as ATCC 1470 modified leptospira medium (Table 13.1). Neomycin may be used to avoid overgrowth in urine cultivation. A series of tubes with 5 ml of medium are inoculated, each with a drop of urine. Two tubes are inoculated with undiluted urine, two tubes with a 1:10 dilution, in PBS, two tubes with a 1:100 dilution and two tubes with a 1:1000 dilution. A 30-μg neomycin antibiogram disc is added into each tube.[8] For examination of the kidney a piece of the kidney is placed in a 5-ml syringe and crushed with the plunger, so a few drops can be inoculated into the medium. The inoculated cultures are kept at room temperature. During the first week nothing is done, but then a drop is examined by dark-field microscopy every week for 5 weeks and then twice per month for 4 months.[6]

Table 13.1 Recipe for the ATCC 1470 Modified Leptospira Medium

Peptone	0.3 g
Beef extract	0.2 g
NaCl	0.5 g
Agar (if necessary)	1.5 g
Distilled water	900.0 ml

Adjust basal medium for a final pH of 7.4, autoclave at 121°C for 15 min., cool to 50°C, and add aseptically:

Sterile rabbit serum	100.0 ml
0.05% Hemin[a]	2.5 ml

Note: Adjust pH to 7.3, ± 0.1, inactivate entire medium for 1 h at 56°C the day following its preparation.

[a] Dissolve hemin in 0.5 to 1.0 ml *N* NaOH and add distilled water to 100 ml. Autoclave at 121°C for 15 min.

Data from ATCC Culture Medium 1470 Modified Leptospira Medium, http://www.atcc.org/catalogs/catalogs.html.

13.2.2.3 Differentiation and identification

The microagglutination test may be used to divide isolates of *L. interrogans* into different serovars. The test may be performed either on a slide or in a microtiter plate using rabbit antisera against the various serotypes. The isolates are passed three times through a liquid medium, e.g., the ATTC medium (Table 13.1). An antigen suspension with an optical density of McFarland 0.5 is prepared. Each antiserum is diluted in two-fold steps starting with 1:25.

Antigen 50 μl and 50 μl of each antibody dilution are mixed in a microtiter plate, shaken gently, and incubated for 4 h. Then agglutination is observed at 450× in a dark-field microscope. The titer is defined as the highest dilution of which at least 50% of the organisms are agglutinated. Further identification requires nucleic acid hybridization studies.[6]

13.2.2.4 Serology

Test for serum antibodies may be performed as described for identification, or as a latex agglutination assay (see Chapter 5). If this is done by latex particles coated with one specific antigen, at least serovar *ballum* and *icterohemorrhagiae* should be included for testing rodents. Multiserovar latex particles are commercially available from Sanofi Pasteur (U.S.).

13.2.3 Safety

The U.S. CDC recommends Biosafety Level-2 practices for all activities involving the use or manipulation of tissues, body fluids, and cultures known or potentially infected with *L. interrogans*, and for the housing of infected animals. Gloves are recommended for the handling and necropsy of infected animals, and when there is the likelihood of direct skin contact with infectious materials. Vaccines are not available for use in humans.[9]

13.3 Spirillum minus

Spirillum minus is a Gram-negative, motile spiral bacterium 3 to 5 μm long causing one form of *rat bite fever* known as *Sodoku*, the symptoms of which are similar to *rat bite fever* caused by *Streptobacillus moniliformis*. *S. minus* cannot be cultivated by any means, and as such it is not possible to isolate it from rats, although it is obvious that rats may harbor this agent. Dark-field microscopy of blood or wound exudates from patients, with a history of rat bites, may reveal the presence of the spiral organisms. The disease has been reported primarily in Japan, but recent cases have also been reported in Kenya[10] and Brazil.[11] Gentamicin or penicillin are both efficient treatment against this agent.

References

1. Ohder, H. and Wullenweber, M., *Treponema cuniculi*, in *Diagnostic Microbiology for Laboratory Animals*, Kunstyr, I., Ed., Gustav Fischer, New York, 1992, 107.
2. Cox, D.L., Riley, B., Chang, P., Sayahtaheri, S., Tavell, S., and Hevelone, J., Effects of molecular oxygen, oxidation-reduction potential, and antioxidants upon in vitro replication of *Treponema pallidum* subsp. *pallidum*, *Appl. Environ. Microbiol.*, p. 3063, 1990.
3. Cunliffe-Beamer, T. L. and Fox, R. R., Venereal spirochetosis of rabbits: description and diagnosis, *Lab. Anim. Sci.*, 31, 366, 1981.

4. Ijsselmuiden, O.E., van der Sluis, J.J., Mulder, A., Stolz, E., Bolton, K.P., and van EijK, R.V.W., An IgM enzyme-linked immunosorbent assay to detect IgM antibodies to treponemes in patients with syphilis, *Genitourin. Med.*, 65, 79, 1989.

5. Wicher, K., Noorhoek, G.T., Abbruscato, F., and Wicher, V., Detection of *Treponema pallidum* in early syphilis by DNA amplification, *J. Clin. Microbiol.*, 30, 497, 1992.

6. Kaufmann, A. F., and Weyant, R. S., Leptospiraceae, in *Manual of Clinical Microbiology*, Murray, P. R., Baron, E. J., Pfaller, M. A., Tenover, F. C., and Yolken, R. H., Eds., ASM Press, Washington, D.C., 1995, 621.

7. Ohder, H. and Wullenweber, M., *Leptospira* sp., in *Diagnostic Microbiology for Laboratory Animals*, Kunstyr, I., Ed., Gustav Fischer, New York, 1992, 107.

8. Sulzer, C. R. and Jones, W. L., Leptospirosis. Methods in Laboratory Diagnosis, CDC Publication No. 74-8275, U.S. Department of Health, Education and Welfare, Washington, D.C., 1978.

9. Office of Health and Safety, Centers for Disease Control and Prevention, Laboratory Biosafety Level Criteria, http://www.cdc.gov/od/ohs/biosfty/bmbl/section3.htm.

10. Bhatt, K. M. and Mirza, N. B., Rat bite fever: a case report of a Kenyan, *East. Afr. Med. J.*, 69(9), 542, 1992.

11. Hinrichsen, S. L., Ferraz, S., Romeiro, M., Muniz Filho, M., Abath, A. H., Magalhaes, C., Damasceno, F., Araujo, C. M., Campos, C. M., and Lamprea, D. P. [Sodoku — a case report], *Rev. Soc. Bras. Med. Trop.*, 25(2), 135, 1992.

12. ATCC Culture Medium 1470 Modified Leptospira Medium, http://www.atcc.org/catalogs/catalogs.html.

chapter fourteen

Mollicutes

Contents

14.1 *Mycoplasma* 223
    14.1.1 Characteristics of infection 223
    14.1.2 Characteristics of the agent 225
        14.1.2.1 Morphology 225
        14.1.2.2 Cultivation 225
        14.1.2.3 Identification 228
        14.1.2.4 Serology 228
14.2 *Acholeplasma* 229
References 229

Members of the class Mollicutes are characterized by the lack of a cell wall and are the smallest self-propagating cells known. Two genera of Mollicutes are of interest in laboratory animal bacteriology: rodents and rabbits may harbor *Mycoplasma*, and guinea pigs and rabbits may additionally harbor *Acholeplasma*. They differ simply in their demands for cholesterol in the medium: *Acholeplasma* grows on cholesterol-free media, and *Mycoplasma* does not.

14.1 Mycoplasma

14.1.1 Characteristics of infection

M. pulmonis is the *Mycoplasma* most frequently encountered in laboratory animal bacteriology. In most individual cases the sole presence of this agent in the host causes either very mild symptoms or no symptoms at all. However, when complicated with other infectious agents, such as *Pasteurella pneumotropica*,[1] various viruses,[2] or CAR bacillus,[3] as well as environmental inducers such as raised ammonia levels,[4] disease symptoms may become more severe. *M. pulmonis* primarily infects rats, in which snuffles, ruffled hair coat, bronchopneumonia, and arthritis, mostly in a mild form, may occur. Additionally, it colonizes genitals of both males[5] and females,[6] and at least in the latter it may affect reproduction. Even in the abscence of clinical symptoms *M. pulmonis* may be hazardous to experiments. It may raise the incidence of respiratory tract tumors,[7] decrease the cellular and humoral immune response,[8] decrease the severity of adjuvant arthritis,[9] and reduce the incidence of diabetes mellitus type 1 in BB rats.[10] The infection is far less common in mice, but symptoms are similar. In extremely rare cases, the agent may be isolated from rabbits and guinea pigs as well.[11]

M. arthritidis may cause arthritis in rats and mice and probably also in rabbits.[12] Infection is mostly asymptomatic,[13] but immunodeficiency,[14] low social rank,[15] and genetics[16] may act as determinants of disease. In mice it may also cause conjunctivitis and uveitis[17] and dermal necrosis. The latter conditions are under influence of the major histocompatibility complex.[18] It may cause increased susceptibility to experimental pyelonephritis in rats[19] and various types of decreased cellular immune response in both rats[20,21] and rabbits.[22,23]

M. neurolyticum may be isolated from the conjunctiva and nasopharynx of rats and mice, in which it probably does not cause any symptoms at all, although some early descriptions indicate that it may cause conjunctivitis.[24,25] It may infect cell cultures, e.g., murine leukemia cell lines,[26] and cell lines used for propagation of other infectious agents. Infection of the cell cultures may complicate the intercerebral passage of viral and protozooan infections in mice[27,28] and lead to the so-called *rolling disease*, i.e., a disease characterized by rollings in the inoculated mice. Typical rollings are along the axis of the animal and of a duration of approximately 20 to 40 s. This is caused by a toxin produced by *M. neurolyticum;* symptoms may also occur after both intravenous and intraperitoneal injection, and they may occur in both rats and mice.[29] Spontaneous infections are not known to occur as a major problem in commercially bred laboratory rats and mice.

M. collis may be isolated from the conjunctiva and nasopharynx of rats and mice. It may be the cause of conjunctivitis in rats, but this is not clearly documented.The infection is not widespread and has mostly been observed in the U.K.[30]

M. muris has been isolated from the vagina of mice.[31] It has only been isolated once in one colony. As it does not grow on ordinary *Mycoplasma* media, it is difficult to know whether isolation is uncommon due to the fact that infection is uncommon or due to the fact that isolation seldom is attempted. The impact on the mice is unknown.

Guinea pigs may harbor *M. caviae*[32] and *M. cavipharyngis*.[33] From hamsters *M. cricetuli*[34] and *M. oxoniensis*[35] may be isolated. The pathogenicity of these organisms seems to be rather low.

14.1.2 Characteristics of the agent

14.1.2.1 Morphology

Mycoplasmas are clearly Gram negative, as they have no cell wall, but since they are not easily stained by Gram staining other methods should be applied. Smears may be prepared for staining by mounting a punched-out colony on a slide with the growing side on the glass, holding it tilted and melting the agar onto the slide in a water bath, air-drying, and fixating with Bouin's solution for 30 min.[36] Such slides may be Giemsa stained. However, the method of choice for studying the cellular morphology seems to be dark-field or phase-contrast microscopy on nonfixed aqueous immersions. This will also allow demonstration of motility. Immunofluorescence staining directly on colony smears is also an option if the identity is known well enough to choose an antibody for staining.

The cellular morphology of mycoplasmas is rather variable, ranging from round to filamentous. The general principle is that young cultures in their logarithmic growth phase contain numerous filaments, while older cultures contain more coccoid forms. *M. pulmonis* is usually spherical to pear shaped and with a diameter of 0.3 to 0.8 μm. Filaments may occur. *M. neurolyticum* has the same shape as *M. pulmonis* and may show filaments up 160 μm long. *M. arthritidis* are pleomorphic, and smaller filaments may occur.

To be able to observe colony morphology agar plates should be inspected under a binary microscope. Colonies may vary in size from 15 μm to more than 300 μm. Typical *Mycoplasma* colonies are said to have a fried-egg appearance, i.e. large colonies with a top in the center, but this is not necessarily a common characteristic for rodent myoplasmas. *M. arthritidis* usually produces such colonies, while *M. pulmonis* and *M. neurolyticum* do not. *M. pulmonis* grows with large and raised colonies with a granulated or vacuolated surface. Other *Mycoplasma* spp. grow with multiple, hardly visible colonies spread all over the agar surface.

14.1.2.2 Cultivation

Isolation may be attempted in broth as well as in solid medium. In general, media for *Mycoplasma*, in addition to the basic substances, contain yeast, horse serum, and penicillin. If *M. neurolyticum* is to be isolated penicillin should, however, be omitted, as this agent is inhibited by the addition of penicillin. Horse serum should be inactivated by heating to 56°C for 30 min before adding to the medium.

The media in Table 14.1. may be used. Both solid media and broth should be heavily inoculated, e.g., with minced tissue, if possible. It should, however, be noted that in glucose-containing broths the tissue may induce a yellow change, which is not related to glucose metabolism. Ordinary media do not support growth of *M. muris*, which should be grown in the SP4 broth or on the SP4 agar.

Table 14.1 Media for Isolation of *Mycoplasma* spp.

	Mycoplasma agar	Mycoplasma broth	SP4 broth	SP4 agar
Mycoplasma agar (BBL, U.S.)	34.0 g			
Mycoplasma broth base (BBL, U.S.)		10.5 g	3.5 g	3.5 g
Noble agar (Difco, U.S.)				35.0 g
Peptone (Difco, U.S.)			5.3 g	5.3 g
Tryptone (Difco, U.S.)			10.0 g	10.0 g
Glucose			5.0 g	5.0 g
Distilled water	680 ml	680 ml	520 ml	520 ml

Mix, boil, and autoclave at 121°C for 15 min.
Adjust pH to 7.2 ± 2

Add the following sterile supplements:

	Mycoplasma agar	Mycoplasma broth	SP4 broth	SP4 agar
CMRL 1066 medium (10×) with glutamine			50 ml	50 ml
Fetal bovine serum	200 ml	200 ml	170 ml	170 ml
Yeast extract solution (Gibco, U.S.)	100 ml	100 ml	140 ml	140 ml
Yeastolate (2% solution) (Difco, U.S.)			100 ml	100 ml
Penicillin	500,000 i.u.	500,000 i.u.		
Thallium acetate	500 mg	500 mg		
Phenol red solution (0.1% solution)	20 ml	20 ml	20 ml	20 ml

Data from Ohder and Wullenweber[36] and Tully et al.[50]

To detect growth phenol red should be added to broths. For those mycoplasmas that ferment glucose 1% glucose should be added, while for those
mycoplasmas that utilize arginine 1% arginine should be added.

The inoculated media are incubated at 37°C with preferably 5% CO_2,
as described for microaerophilic incubation (see Chapter 3), for example.
Additionally, a humid atmosphere should be used for cultivation on solid
media. Growth will usually be achieved in 6 to 10 days, but occasionally
solid media have to be incubated for up to 3 weeks. Broths should not
be incubated more than 1 week, after which they should be plated on a
solid medium.

Some optimal sampling sites for various mycoplasmas are listed in
Tables 14.2 and 14.3.

Table 14.2 Characteristics of *Mycoplasma* spp. from Rats and Mice

Mycoplasma	Species	Isolation site	Growth on serum yeast Mycoplasma agar	Penicillin resistant	Glucose catabolism	Utilizes arginine	Hem-adsorption
M. pulmonis	Mice, rats, (rabbits, guinea pigs in rare cases)	Respiratory system, genitals, joints	+	+	+	–	d
M. neurolyticum	Mice, rats	Conjunctiva, nasopharynx	+	–	+	–	–
M. muris	Mice	Vagina	–	+	–	+	+
M. arthritidis	Mice, rats, rabbits	Joints, conjunctiva, skin	+	+	–	+	–
M. collis	Mice, rats	Conjunctiva, Harderian gland, nasopharynx	+	+	+	–	?

Table 14.3 *Mycoplasma* spp. Found in Rabbits, Guinea Pigs, and Hamsters

Mycoplasma	Species	Isolation site	Glucose catabolism	Utilizes arginine	Hemadsorption
M. caviae	Guinea pigs	Nasopharynx, vagina	+	d	–
M. caviapharyngis	Guinea pigs	Nasopharynx	+	–	+
M. cricetuli	Hamsters	Conjunctiva	+	–	–
M. oxoniensis	Hamsters	Conjunctiva	+	–	–

14.1.2.3 Identification

The ability to adsorb erythrocytes is characteristic for some *Mycoplasma* spp. (Table 14.2). Guinea pig erythrocytes should be used as standard, but the species from which the isolate originated should also be used, possibly along with erythrocytes from a third species.

In many laboratories identification mostly has been based on immunological assays using either monoclonal- or polyclonal-specific antibodies. A widely used method is inhibition of growth on an agar plate by placing a paper disk soaped in a dilution of a high-titer antiserum in the middle of the plate.[37] The dilution to be used must be found in pretesting with a sample of known identity. Immunofluorescence staining, dot blotting, and Western immunoblotting using a range of specific antimycoplasmal or anti-acholeplasmal sera for identification may be performed as described in Chapter 5. Blotting may detect approximately down to 10^4 colony-forming units per milliliter depending on mycoplasma species.[38] Cross-reaction between closely related *Mycoplasma* spp., e.g., *M. pulmonis* and *M. arthritidis*, may be problematic in all immunological *Mycoplasma* identification assays depending on the nature of the antibody used.

Several gene sequences are known for rodent mycoplasmas, and, therefore, molecular biological methods may be applied for identification and detection purposes. Exceptions are *M. cavipharyngis*, *M. cricetuli*, and *M. oxoniensis*, which thus far have not been sequenced. Restriction endonuclease analysis of chromosomal DNA as described in Chapter 5 may be used for identifying isolates. Generally, the results of restriction analysis correspond well to antigenic similarities, although it has been shown that at least for *M. arthritidis* it does not allow grouping with as fine a precision as immunoblotting.[39] Polymerase chain reaction (PCR) (see Chapter 5) is an optimal way of identifying mycoplasmas. Highly sensitive RT-PCR assays have been developed, in which the nucleic acid content equivalent to a single organism may be detected. Primers species specific for *M. pulmonis*, *M. arthritidis*, *M. neurolyticum*, *M. muris*, and *M. collis* may be selected, while genus-specific primers may react with all mycoplasmal species as well as with members of the genera *Ureaplasma*, *Spiroplasma*, and *Acholeplasma* without cross-reaction with members of the genera *Streptococcus*, *Lactobacillus*, *Bacillus*, and *Clostridium*.[40]

PCR may also be used directly to detect *Mycoplasma* spp. in tissue samples. Tissue samples are frozen in liquid nitrogen. Approximately 1 g of the frozen tissue is ground into a fine powder and placed in a sterile glass with 1 ml of 0.1 *M* NaCl, 0.1 *M* Tris-HCl, and 1% SDS (pH 8.0), boiled for 10 min, and spun at 10.000 g for 3 min. The supernatant is used for RT-PCR. As little as 1 pg of nucleic acid may be detected.[41]

14.1.2.4 Serology

Serology has been performed to detect infection with *M. pulmonis* in rodent colonies for several years. The first assay to be applied for this purpose was the complement fixation assay, which, however, is too insensitive to be of any practical use today. Today the method of choice is either immunofluorescence assay[42] or ELISA[43] (see Chapter 5). ELISA has also been widely used for detection of antibodies against *M. arthritidis*. These two mycoplasmas generally cross-react depending on the test applied.[44] If both agents are to be included in a health-monitoring profile, specific ELISA assays with both *M. pulmonis* and *M. arthritidis* antigen should be run to secure a high sensitivity for both infections. Tests for IgG are more sensitive than tests for IgM, but generally serological assays are fully reliable, and for spontaneous infections they seem to be more sensitive than cultivation methods.[45] Antigens are commercially available from Harlan (U.K.), Charles River (U.S.), and Organon Teknika (U.S.).

14.2 Acholeplasma

Guinea pigs and rabbits may harbor various *Acholeplasma*[46] species. These organisms do not require cholesterol in the medium. They may grow aerobically, glucose is metabolized by some, and arginine and urea are not hydrolyzed. In guinea pigs at least *A. cavigenitalium* and *A. laidlawii*[47] have been identified. Spontaneous antibodies to *A. laidlawii* membrane lipids may be found in guinea pig sera. *A. laidlawii*[48] and *A. multilocale*[49] have been isolated from the feces of rabbits. The pathogenicity of these organisms seems to be rather low.

Cultivation and identification of these agents are similar to methods described for mycoplasmas. Optimal sampling sites may be found in Table 14.4.

Table 14.4 Acholeplasma spp. Found in Rabbits, Guinea Pigs, and Hamsters

Acholeplasma	Species	Isolation site
A. cavigenitalium	Guinea pigs	Vagina
A. laidlawii	Rabbits, guinea pigs	Feces, vagina, nasopharynx
A. multilocale	Rabbits	Feces

References

1. Brennan, P. C., Fritz, T. E., and Flynn, R. J., Role of *Pasteurella pneumotropica* and *Mycoplasma pulmonis* in murine pneumonia, *J. Bacteriol.*, 97, 337, 1969.
2. Schoeb, T. R., Kervin, K. C., and Lindsey, J. R., Exacerbation of murine respiratory mycoplasmosis in gnotobiotic F344/N rats by Sendai virus infection, *Vet. Pathol.*, 22, 272, 1985.

3. Waggie, K., Kagiyama, N., and Itoh, T., "Cilia Associated Respiratory" (CAR) Bacillus, in *Manual of Microbiologic Monitoring of Laboratory Animals*, NIH Publication No. 94-2498, Waggie, K., Kagiyama, N., Allen, A. M., and Nomura, T., Eds., U.S. Department of Health and Human Services, Public Health Service, National Institutes of Health, National Center for Research Resources, Bethesda, MD, 1994, 121.

4. Broderson, J. R., Lindsey, J. R., and Crawford, J. E., The role of environmental ammonia in respiratory mycoplasmosis of rats, *Am. J. Path.*, 85, 115, 1976.

5. Juhr, N. C., Ratsch, H., Stuckenberg, R., and Schiller, S., Colonization of *Mycoplasma pulmonis* to the genital organ of male rats, *Tierlaboratorium*, 12, 224, 1989.

6. Casillo, S. and Blackmore, D. K., Uterine infections caused by bacteria and mycoplasmas in mice and rats, *J. Comp. Path.*, 82, 477, 1972.

7. Kimbrough, R. and Gaines, T. B., Toxicity of hexylmethylphosphoramide in rats, *Nature*, 211, 146, 1966.

8. Lai, W. C., Pakes, S. P., Owusu, I., and Wang, S., *Mycoplasma pulmonis* depresses humoral and cell-mediated responses in mice, *Lab. Anim. Sci.*, 39(1), 11, 1989.

9. Taurog, J. D., Leary, S. L., Cremer, M. A., Mahowald, M. L., Sandberg, G. P., and Manning, P. J., Infection with *Mycoplasma pulmonis* modulates adjuvant- and collagen-induced arthritis in Lewis rats, *Arth. Rheum.*, 27, 943, 1984.

10. Kloeting, I., Sadewasser, S., Lucke, S., Vogt, L., and Hahn, H. J., Development of BB rat diabetes is delayed or prevented by infections or applications of immunogens, in *Frontiers in Diabetes Research*, Shafrir, E. and Renolds, A. E., Eds., John Libbey, London, 1988, 190.

11. Lindsey, J. R., Cassell, G. H., Davis, J. K., and Davidson, M. K., Mycoplasmal and other bacterial diseases of the respiratory system, in *The Mouse in Biomedical Research Vol. II*, Foster, H. L., Small, J. D., and Fox, J. G., Eds., Academic Press, New York, 21, 1982.

12. Washburn, L. R., Cole, B. C., Gelman, M. I., and Ward, J. R., Chronic arthritis of rabbits induced by mycoplasmas. I. Clinical microbiologic, and histologic features, *Arth. Rheum.*, 23(7), 825, 1980.

13. Cox, N. R., Davidson, M. K., Davis, J. K., Lindsey, J. R., and Cassell, G. H., Natural mycoplasmal infections in isolator-maintained LEW/Tru rats, *Lab. Anim. Sci.*, 38(4), 381, 1988.

14. Binder, A., Hedrich, H. J., Wonigeit, K., and Kirchhoff, H. J., The *Mycoplasma arthritidis* infection in congenitally athymic nude rats, *J. Exp. Anim. Sci.*, 35(4), 177, 1993.

15. Gärtner, K., Kirchhoff, H., Mensing, K., and Velleuer, R., The influence of social rank on the susceptibility of rats to *Mycoplasma arthritidis*, *J. Behav. Med.*, 12(5), 487, 1989.

16. Binder, A., Gärtner, K., Hedrich, H. J., Hermanns, W., Kirchhoff, H., and Wonigeit, K., Strain differences in sensitivity of rats to *Mycoplasma arthritidis* infection are under multiple gene control, *Infect. Immun.*, 58(6), 1584, 1990.

17. Thirkill, C. E. and Gregerson, D. S., *Mycoplasma arthritidis*-induced ocular inflammatory disease, *Infect. Immun.*, 36(2), 775, 1982.

18. Cole, B. C., Piepkorn, M. W., and Wright, E.C., Influence of genes of the major histocompatibility complex on ulcerative dermal necrosis induced in mice by *Mycoplasma arthritidis*, *J. Invest. Dermatol.*, 85(4), 357, 1985.

19. Thomsen, A. C. and Rosendal, S., Mycoplasmosis — experimental pyelonephritis in rats, *APMIS*, 82, 94, 1974.

20. Simberkoff, M. S., Thorbecke, G. J., and Thomas, L., Studies on PPLO infection. Inhibition of lymphocyte mitosis and antibody formation by mycoplasmal extracts, *J. Exp. Med.*, 129, 1163, 1969.
21. Specter, S. C., Bendinelli, M., Ceglowski, W. S., and Friedman, H., Macrophage-induced reversal of immunosuppression by leukemia viruses, *Fed. Proc.*, 37, 97, 1978.
22. Berquist, L. M., Lav, B. H. S., and Winter, C. E., Mycoplasma-associated immuno-suppression: effect on hemagglutinin response to common antigens in rabbits, *Infect. Immun.*, 9, 410, 1974.
23. Westerberg, S. C., Smith, C. B., Wiley, B. B., and Jensen, C., Mycoplasma-virus interrelationships in mouse tracheal organ cultures, *Infect. Immun.*, 5, 840, 1972.
24. Nelson, J. B., Association of a special strain of pleuropneumonia-like organisms with conjunctivitis in a mouse colony, *J. Exp. Med.*, 91, 309, 1950.
25. Nelson, J. B., The relation of pleuropneumonia-like organisms to the conjunctival changes occurring in mice of the Princeton strain, *J. Exp. Med.*, 92, 431, 1950.
26. Tully, J. G. and Rask-Nielsen, R., *Mycoplasma* in leukemic and non-leukemic mice, *Ann. N.Y. Acad. Sci.*, 143, 345, 1967.
27. Sabin, A. B., Isolation of a filterable, transmissible agent with "neurolytic" properties from Toxoplasma infected tissues, *Science*, 88, 189, 1938.
28. Sabin, A. B., Identification of the filterable, transmissible agent with neurolytic agent from Toxoplasma infected tissues as a new pleuropneumonia-like microbe, *Science*, 88, 575, 1938.
29. Committee on Infectious Diseases of Mice and Rats, *Infectious Diseases of Mice and Rats*, National Academy Press, Washington, D.C., 1991, 78.
30. Tully, J. G., Biology of rodent mycoplasmas, in *Viral and Mycoplasmal Infections of Laboratory Rodents, Effects on Biomedical Research*, Bhatt, P. N., Jacoby, R. O., Morse, H. C., III, and New, A. E., Eds., Academic Press, Orlando, 1986, chap. 7.
31. McGarrity, G. J., Rose, D. L., Kwiatkowski, V., Dion, A. S., Phillips, D. M., and Tully, J. G., *Mycoplasma muris*, a new species from laboratory mice, *Int. J. Syst. Bacteriol.*, 33, 350, 1983.
32. Tully, J. G., Biology of rodent mycoplasmas, in *Viral and Mycoplasmal Infections of Laboratory Rodents, Effects on Biomedical Research*, Bhatt, P. N., Jacoby, R. O., Morse, H. C., III, and New, A. E., Eds., Academic Press, Orlando, 1986, chap.7.
33. Hill, A. C., *Mycoplasma cavipharyngis*, a new species isolated from the nasopharynx of guinea pigs, *J. Gen. Microbiol.*, 130, 3183, 1984.
34. Hill, A. C., *Acholeplasma cavigenitalium* sp. nov., isolated from the vagina of guinea pigs, *Int. J. Syst. Bacteriol.*, 42(4), 589, 1992.
35. Hill, A. C., *Mycoplasma oxoniensis*, a new species isolated from Chinese hamster conjunctivas, *Int. J. Syst. Bacteriol.*, 41(1), 21, 1991.
36. Ohder, H. and Wullenweber, M., *Mycoplasma* spp., in *Diagnostic Microbiology for Laboratory Animals*, Kunstyr, I., Ed., Gustav Fischer, New York, 1992, 85.
37. Clyde, W. A., *Mycoplasma* species identification based upon growth inhibition by specific antisera, *J. Immunol.*, 92, 958, 1964.
38. Kotani, H. and McGarrity, G. J., Rapid and simple identification of mycoplasmas by immunobinding, *J. Immunol. Methods*, 85(2), 257, 1985.
39. Washburn, L. R., Voelker, L. L., Ehle, L. J., Hirsch, S., Dutenhofer, C., Olson, K., and Beck, B., Comparison of *Mycoplasma arthritidis* strains by enzyme-linked immunosorbent assay, immunoblotting, and DNA restriction analysis, *J. Clin. Microbiol.*, 33(9), 2271, 1995.

40. van Kuppeveld, F. J., van der Logt, J. T., Angulo, A. F., van Zoest, M. J., Quint, W. G., Niesters, H. G., Galama, J. M., and Melchers, W. J., Genus- and species-specific identification of mycoplasmas by 16S rRNA amplification, *Appl. Environ. Microbiol.*, 58(8), 2606, 1992.

41. Sanchez, S., Tyler, K., Rozengurt, N., and Lida, J., Comparison of a PCR- based diagnostic assay for *Mycoplasma pulmonis* with traditional detection techniques, *Lab. Anim.*, 28, 249, 1994.

42. Kraft, V., Meyer, B., Thunert, A., Deerberg, F., and Rehm, S., Diagnosis of *Mycoplasma pulmonis* infection of rats by an indirect immunofluorescence test compared with 4 other diagnostic methods, *Lab. Anim.*, 16(4), 369, 1982.

43. Cassell, G. H., Lindsey, J. R., Davis, J. K., Davidson, M. K., Brown, M. B., and Mayo, J. G., Detection of natural *Mycoplasma pulmonis* infection in rats and mice by an enzyme linked immunosorbent assay (ELISA), *Lab. Anim. Sci.*, 31(6), 676, 1981.

44. Minion, F. C., Brown, M. B., and Cassell, G. H., Identification of cross-reactive antigens between *Mycoplasma pulmonis* and *Mycoplasma arthritidis*, *Infect. Immun.* 43(1), 115, 1984.

45. Davidson, M. K., Lindsey, J. R., Brown, M. B., Schoeb, T. R., and Cassell, G. H., Comparison of methods for detection of *Mycoplasma pulmonis* in experimentally and naturally infected rats, *J. Clin. Microbiol.*, 14(6), 646, 1981.

46. Stalheim, O. H. and Matthews, P. J., Mycoplasmosis in specific-pathogen-free and conventional guinea pigs, *Lab. Anim. Sci.*, 25(1), 70, 1975.

47. Hill, A., Isolation of *Acholeplasma laidlawii* from guinea pigs, *Vet. Rec.*, 94(17), 385, 1974.

48. Angulo, A. F., Doeksen, M., Hill, A., and Polak-Vogelzang, A. A., Isolation of Acholeplasmatales from rabbit faeces, *Lab. Anim.*, 21(3), 201, 1987.

49. Hill, A. C., Polak-Vogelzang, A. A., and Angulo, A. F., *Acholeplasma multilocale* sp. nov., isolated from a horse and a rabbit, *Int. J. Syst. Bacteriol.*, 42(4), 513, 1992.

50. Tully, J. G., Whitcomb, R. F., Clark, H. F., and Williamson, D. L., Pathogenic mycoplasmas: cultivation and vertebrate pathogenicity of a new *Spiroplasma*, *Science*, 195, 892, 1977.

Appendix I

Producers of reagents for laboratory animal bacteriology

Some producers have agents, branches, and offices worldwide. For those companies with Internet homepages, it is recommended to use this means of obtaining information on local distributors.

Animal Plant and Health Inspection Services
Center for Animal Health Monitoring (CAHM)
att. NAHMS
555 South Howes
Fort Collins, CO 80521
Tel: +1 (970) 490-8000
Fax: +1 (970) 490-7899
E-mail: NAHMSweb@usda.gov
http://www.aphis.usda.gov/vs/ceah/cahm

BBL
see Becton Dickinson Europe

Becton Dickinson Europe
5 Chemin des Sources
BP 37
F-38241 Meylan CEDEX
France
Tel: +33 476 416464
Fax: +33 476 41856
http://www.bdms.com
Note: BBL and Difco are available from the U.S. office of Becton Dickinson.

bioMérieux SA
F-69280 Marcy l'étoile
France
Tel: +33 478 87 20 00
Fax: +33 478 87 20 90
E-mail: corinne_daubignard@ml.biomerieux.fr
http://www.biomerieux.fr

Charles River Laboratories, Inc.
251 Ballardvale Street
Wilmington, MA 01887-1000
Tel: +1 (800) 522-7287/+1 (978) 658-6000
Fax: +1 (800) 992-7329/+1 (978) 658-7132
E-mail: comments@criver.com
http://www.criver.com

DAKO Corporation
6392 Via Real
Carpinteria, CA 93013
Tel: +1 (805) 566-6655
Fax: +1 (805) 566-6688
E-mail: general@dakousa.com
http://www.dako.com/dakocomp

Difco
See Becton Dickinson Europe

Gen-Probe Incorporated
10210 Genetic Center Drive
San Diego, CA 92121-1589
Tel: +1 (619) 410-8000
Fax: +1 (619) 410-8001
E-mail cindyi@gen-probe.com
http://www.gen-probe.com

Gibco
Life Technologies, Inc.
9800 Medical Center Drive
P.O. Box 6482
Rockville, MD 20849-6482
Tel: +1 (800) 338-5772
http://www.lifetech.com

Harlan UK Limited
Shaw's Farm
Blackthorn, Bicester, Oxon
GB-0X6 OTP
United Kingdom
Tel.: +44 1869 243241
Fax: +44 1869 246759
http://www.harlan.com

ICN Biomedical Research Products
3300 Hyland Avenue
Costa Mesa, CA 92626
Tel: +1 (714) 545-0100
Fax: +1 (714) 557-4872
http://www.icnbiomed.com

Lee Laboratories
1475 Athens Hwy
Grayson, GA 30017
Tel: +1 (800) 732-9150/+1 (770) 972-4450
Fax: +1 (770) 979-9570
E-mail Sales@LeeLabs.com
http://www.leelabs.com

Merck KgaA
Frankfurterstr. 250
D-64293 Darmstadt
Germany
Tel.: +49 61 51 72 0
Fax: +49 61 51 72 2000
E-mail: service@merck.de
http://www.merck.de

Meridian Diagnostics Inc.
3471 River Hills Drive
Cincinnati, OH 45244
Tel: +1 (513) 271-3700/+1 (800) 543-1980
Fax: +1 (513) 271-0124

New Brunswick Scientific Co., Inc.
44 Talmadge Road
Edison, NJ 08818-4005
Tel: +1 (732) 287-1200/+1 (800) 631-5417
Fax: +1 (732) 287-4222
E-mail: bioinfo@nbsc.com
http://www.nbsc.com

New England Biolabs, Inc.
32 Tozer Road
Beverly, MA 01915
Email: info@neb.com
http://www.neb.com

Organon Teknika Corporation
100 Akzo Avenue
Durham, NC 27712
Tel: +1 (919) 620-2000/ +1 (800) 682-2666
Fax: +1 (800) 432-9682
E-mail: customerservice@orgtek.com
http://www.organonteknika.com

Oxoid Ltd.
Wade Road
Basingstoke, Hampshire
RG24 8PW
United Kingdom
Tel: +44 0 1256 841144
Fax: +44 0 1256 463388

Perkin-Elmer Corporation
761 Main Avenue
Norwalk, CT 06859-0001
Tel: +1 (203) 762-1000/+1 (800) 762-4000
Fax: +1 (203) 762-6000
E-mail: info@perkin-elmer.com
http://www.perkin-elmer.com

Rosco Ltd.
Taastrupgaardsvej 30
DK-2630 Taastrup
Denmark
Tel: +45 43 99 33 77
Fax: +45 42 52 73 74
Email: info@as-rosco.dk
http://www.rosco.dk

Sanofi Pasteur Diagnostics, Inc.
1000 Lake Hazeltine Drive
Chaska, MN 55318
Tel: +1 (612) 448-4848
Fax: +1 (612) 368-1110
http://www.mbbnet.umn.edu/company_folder/sdp.html

Sigma
3050 Spruce Street
P.O. Box 14508
St. Louis, MO 63178
Tel: +1 (314) 771-5765
Fax: +1 (314) 771-5757
E-mail sigma@sial.com
http://www.sigma-aldrich.com

Stratagene Inc.
11011 North Torrey Pines Road
La Jolla, CA 92037
Tel: +1 (800) 894-1304/+1 (619) 535-5400
Fax: +1 (619) 535-0071
E-mail: techsvc@stratagene.com
http://www.stratagene.com

Appendix II

Biosafety levels for microbiological laboratories

Bacteria posing a risk to laboratory staff should be handled with certain precautions. The U.S. Centers for Disease Control and Prevention (CDC)[1] has set up four biosafety levels for work with microorganisms. These criteria should be consulted for work performed outside U.S. borders as well. The information in this appendix is, in principle, a short review of the CDC recommendations.*

It should be noted that different procedures for the same agent may require different biosafety levels. In general, procedures are divided into three different types:

- The use or manipulation of known or potentially infectious tissues, body fluids, and cultures
- Housing and handling infected animals
- Work involving production volumes or concentrations of cultures

Biosafety Level 1 (Table A.II.1) is sufficient for work involving well-characterized agents not known to cause disease in healthy adult humans and of minimal potential hazard to laboratory personnel and the environment. The laboratory is not necessarily separated from the general traffic patterns in the building. Work is generally conducted on open bench tops using standard microbiological practices. Special containment equipment or facility design is not required or generally used. Laboratory personnel have specific training in procedures conducted in the laboratory and are supervised by a scientist with general training in microbiology or a related science.[2]

Biosafety Level 1 covers the majority of the procedures and the agents discussed in this book. Biosafety Level 1 is insufficient for work involving *Campylobacter* spp., *Leptospira* spp., *Mycobacterium* spp., *Salmonella* spp., and *Francisella tularensis*.

* Office of Health and Safety, Centers for Disease Control and Prevention, 1600 Clifton Road N.E., Mail Stop F05 Atlanta, GA 30333.

Table A.II.1 Standard and Special Safety Practices, Equipment, and Facilities for Work with Agents Assigned to Biosafety Level 1

A. *Standard Microbiological Practices*
   1. Limited access to laboratory.
   2. Handwashing after handling viable materials and animals, after removing gloves, and before leaving the laboratory.
   3. No eating, drinking, smoking, handling contact lenses, or applying cosmetics in the laboratory. Persons who wear contact lenses in laboratories are to wear goggles or a face shield. Food to be stored outside the work area in cabinets or refrigerators designated and used for this purpose only.
   4. No mouth pipetting.
   5. Avoid creating splashes or aerosols.
   6. Daily decontamination of work and after any spill of viable material.
   7. Decontamination of all cultures, stocks, and other regulated wastes before disposal by an approved decontamination method, such as autoclaving. Materials for decontamination outside of the immediate laboratory are to be placed in a durable leakproof container and closed for transport from the laboratory. Materials for decontamination off-site from the laboratory to be packaged in accordance with applicable local, state, and federal regulations, before removal from the facility.
   8. Insect and rodent control program in effect.
B. *Special Practices*: None
C. *Safety Equipment* (Primary Barriers)
   1. Laboratory coats, gowns, or uniforms to prevent contamination or soiling of street clothes.
   2. Gloves, if the skin on the hands is broken or if a rash exists.
   3. Protective eyewear for anticipated splashes of microorganisms or other hazardous materials to the face.
D. *Laboratory Facilities* (Secondary Barriers)
   1. A sink for handwashing in each laboratory.
   2. Easy cleaning design.
   3. Bench tops impervious to water and resistant to acids, alkalis, organic solvents, and moderate heat.
   4. Sturdy laboratory furniture. Spaces between benches and cabinets, and equipment accessible for cleaning.
   5. Fly screens on all windows to open (if any).

From Office of Health and Safety, Centers for Disease Control and Prevention, Laboratory Biosafety Level Criteria.[2]

Biosafety Level 2 (Table A.II.2) standards are similar to Level 1 and suitable for work involving agents of moderate potential hazard to personnel and the environment. Biosafety Level-2 criteria differ from Level 1 as follows:

Table A.II.2 Standard and Special Safety Practices, Equipment, and Facilities for Work with Agents Assigned to Biosafety Level 2

A. *Standard Microbiological Practices*: As for Biosafety Level 1
B. *Special Practices*
   1. No persons who are at increased risk of acquiring infection or for whom infection may be unusually hazardous.
   2. Only persons advised of the potential hazard and who meet specific entry requirements.
   3. Special provisions for entry, and a hazard warning sign (the universal biohazard symbol, name of the infectious agent, name and telephone number of the responsible person(s), and the special requirement(s) for entering the laboratory) on the access door.
   4. Staff immunizations or tests for the agents handled or potentially present in the laboratory.
   5. Possibly serum samples from laboratory and other at-risk personnel.
   6. A biosafety manual.
   7. Appropriate staff training including annual updates on potential hazards, necessary precautions, and exposure evaluation procedures.
   8. Precaution with any contaminated sharp items, including needles and syringes, slides, pipettes, capillary tubes, and scalpels. Plasticware instead of glassware whenever possible.
   9. Cultures, tissues, or specimens of body fluids to be placed in a container that prevents leakage during collection, handling, processing, storage, transport, or shipping.
   10. Laboratory equipment and work surfaces to be decontaminated with an appropriate disinfectant on a routine basis, after work with infectious materials, and especially after overt spills, splashes, or other contamination by infectious materials. Contaminated equipment to be decontaminated before removal from the facility.
   11. Major spills and accidents to be reported immediately to the laboratory director.
   12. No animals other than those involved in the work.
C. *Safety Equipment* (Primary Barriers)
   1. Properly maintained biological safety cabinets (BSC), preferably Class II, or other appropriate personal protective equipment or physical containment devices to be used whenever:
      a. Procedures with a potential for creating infectious aerosols or splashes are conducted.
      b. High concentrations or large volumes of infectious agents are used.
   2. Face protection to be used for anticipated splashes or sprays of infectious or other hazardous materials to the face, when the microorganisms must be manipulated outside BSC.

continued

Table A.II.2 (continued) Standard and Special Safety Practices, Equipment, and Facilities for Work with Agents Assigned to Biosafety Level 2

---

   3. Protective laboratory coats, gowns, smocks, or uniforms designated for lab use to be worn while in laboratory and removed and left in the laboratory before leaving.
   4. Gloves to be worn when handling infected animals and when hands may contact infectious materials, contaminated surfaces, or equipment.
D. **Laboratory Facilities** (Secondary Barriers): As for Biosafety Level 1 with the following addition:
   A method for decontamination of infectious or regulated laboratory wastes and an eyewash facility to be available.

From Office of Health and Safety, Centers for Disease Control and Prevention, Laboratory Biosafety Level Criteria.[2]

- Laboratory personnel have specific training in handling pathogenic agents and are directed by competent scientists
- Access to the laboratory is limited when work is being conducted
- Extreme precautions are taken with contaminated sharp items
- Certain procedures in which infectious aerosols or splashes may be created are conducted in biological safety cabinets or other physical containment equipment[2]

Biosafety Level 2 may be sufficient for all work with *Campylobacter* spp., *Leptospira* spp., *Mycobacterium* spp. (*M. tuberculosis* and *M. bovis* excepted), and *Salmonella* spp., as well as for activities with clinical materials containing or potentially containing *Francisella tularensis*. Biosafety Level-2 practices may also be sufficient for preparation of acid-fast smears, and culturing of clinical specimens containing or potentially containing *M. tuberculosis*, or *M. bovis*, provided that aerosol-generating manipulations of such specimens are conducted in a Class I or II biological safety cabinet.

Biosafety Level 3 (Table A.II.3) is applicable to clinical, diagnostic, teaching, research, or production facilities where work is done with indigenous or exotic agents that may cause serious or potentially lethal disease as a result of exposure by the inhalation route. Laboratory personnel have specific training in handling pathogenic and potentially lethal agents and are supervised by competent scientists who are experienced in working with these agents. All procedures involving the manipulation of infectious materials are conducted within biological safety cabinets or other physical containment devices, or by personnel wearing appropriate personal protective clothing and equipment. The laboratory has special engineering and design features.

For those existing facilities that do not have all the facility safeguards recommended for Biosafety Level 3 (e.g., access zone, sealed penetrations,

Table A.II.3 Standard and Special Safety Practices, Equipment, and
Facilities for Work with Agents Assigned to Biosafety Level 3

A. *Standard Microbiological Practices*: As for Biosafety Level 1.
B. *Special Practices*: As for Biosafety Level 2 with the following additions:
  1. All personnel to demonstrate proficiency in standard microbiological
     practices and techniques, and in the practices and operations specific
     to the laboratory facility.
  2. All manipulations involving infectious materials to be conducted in
     biological safety cabinets or other physical containment devices
     within the containment module. No work in open vessels to be
     conducted on the open bench.
  3. Spills of infectious materials to be decontaminated, contained, and
     cleaned up by appropriate professional staff.
C. *Safety Equipment* (Primary Barriers)
  1. Properly maintained biological safety cabinets (BSC) (Class II or III)
     to be used for all manipulation of infectious materials.
  2. Outside a BSC, appropriate combinations of personal protective
     equipment in combination with physical containment devices to be
     used for manipulations of any material (live or dead) that may
     possibly be a source of infectious aerosols and for sampling and
     necropsing infected animals or embryonated eggs.
  3. Face protection to be worn for manipulation of infectious materials
     outside BSC.
  4. Respiratory protection to be worn when aerosols cannot be safely
     contained (i.e., outside BSC), and in rooms containing infected
     animals.
  5. Protective laboratory clothing to be worn in, and not worn outside,
     the laboratory.
  6. Gloves to be worn when handling infected animals and when hands
     may contact infectious materials and contaminated surfaces or
     equipment.
  7. Reusable laboratory clothing to be decontaminated before being
     laundered. Disposable gloves to be discarded when contaminated,
     and never washed for reuse.
D. *Laboratory Facilities* (Secondary Barriers): As for Biosafety Level 2 with
   the following additions:
  1. The laboratory to be separated from areas which are open to
     unrestricted traffic flow within the building.
  2. Passage into the laboratory through two sets of self-closing doors
     and possibly a clothes-changing room (shower optional).
  3. A ducted exhaust air ventilation system based on a directional airfoil
     that draws air from "clean" areas into the laboratory toward
     "contaminated" areas, no recirculation to any other area, and
     discharge to the outside with filtration and other treatment optional.
     The outside exhaust to be dispersed away from occupied areas and
     air intakes.
  4. An eyewash facility.

continued

Table A.II.3 (continued) Standard and Special Safety Practices, Equipment, and Facilities for Work with Agents Assigned to Biosafety Level 3

---

5. The High Efficiency Particulate Air (HEPA)-filtered exhaust air from BSC to be discharged directly to the outside or through the building exhaust system.
6. Equipment that may produce aerosols to be contained in devices that exhaust air through HEPA filters before discharge into the laboratory.
7. Vacuum lines to be protected with liquid disinfectant traps and HEPA filters, or their equivalent.

From Office of Health and Safety, Centers for Disease Control and Prevention, Laboratory Biosafety Level Criteria.[2]

directional airflow, etc.), acceptable safety may be achieved for routing or repetitive operations (e.g., diagnostic procedures involving the propagation of an agent for identification, typing, and susceptibility testing) in Biosafety Level-2 facilities. However, the recommended standard microbiological practices, special practices, and safety equipment for Biosafety Level 3 must be rigorously followed. The decision to implement this modification of Biosafety Level-3 recommendations should be made only by the laboratory director.[2]

Biosafety Level-3 and Animal Biosafety Level-3 practices, containment equipment, and facilities are recommended for all manipulations of cultures and for experimental animal studies involving *F. tularensis* and for the propagation and manipulation of cultures of *M. tuberculosis* or *M. bovis*.

Biosafety Level 4 is required for work with dangerous and exotic agents which pose a high individual risk of aerosol-transmitted laboratory infections and life-threatening disease. Agents with a close or identical antigenic relationship to Biosafety Level-4 agents are handled at this level until sufficient data are obtained either to confirm continued work at this level, or to work with them at a lower level.[2] Biosafety Level 4 is not mandatory for any agent listed in this book and, therefore, is not examined in further detail here.

References

1. U.S. Department of Health and Human Services, Public Health Service, Centers for Disease Control and Prevention and National Institutes of Health, Biosafety in Microbiological and Biomedical Laboratories, HHS Publication No. (CDC) 93-8395, 3rd ed., U.S. Government Printing Office, Washington, D.C., 1993.
2. Office of Health and Safety, Centers for Disease Control and Prevention, Laboratory Biosafety Level Criteria, http://www.cdc.gov/od/ohs/biosfty/bmbl/section3.htm

Index

α-hemolysis 121, 133, 200
β-galactosidase 69, 94, 96
β-hemolysis 121, 124, 197
β-lactams 143
β-xylosidase 69
γ-hemolysis 121
2,2'-azino-di(3-ethyl-benzthizoline)
      sulfonate-6-diammonium salt 96
2,4-diamino-6,7-diisopropylpteridine 185
3'-diaminobenzidine tetrahydrochloride 83,
    84
3-amino-9-ethylcarbazole 83, 212
3T3 cell line 159

A

A/JCr mice 209
Abdomen 31, 32, 34, 35, 132, 140
Abortions 132, 168, 184
Abscesses 4, 11, 116, 120, 137, 177, 179, 184,
    192
Acepromazine® 19
Acetoin formation 69
Acetylpromazine 19
Acholeplasma 223, 228, 229
    A. cavigenitalium 229
    A. laidlawii 229
    A. multilocale 229
Acid-fast staining 64, 66
Acidification of the drinking water 2
Acinetobacter 59, 198, 199
    A. junii 198
    A. lwoffii 198
Actinobacillus 59, 177, 179
    A. muris 178–180
    A. urea 180
Actinomyces 56, 129, 140
    A. bovis 140
    A. israelii 141
    A. pyogenes 140
Actinomycosis 140
Aerobic cultivation 43–45, 49, 131, 138, 170,
    178, 181, 197
Aerobic growth 55, 62, 115, 132
Aerococcus 114, 115, 118, 123
    A. viridans 123

Aeromonas 166, 185
    A. caviae 185
    A. hydrophila 185
    A. veronii subsp. *sobria* 185
Agar diffusion inhibition assay 65, 67, 185
Agarose gel electrophoresis 110
Age 2, 4, 6, 7, 177, 211
Agent 3
Agglutination 43, 76, 176, 181, 197, 198, 200,
    220
    latex agglutination assay 121, 147, 220
    hemagglutination 217
    microagglutination 219
    tube agglutination 200
Agrobacterium 59, 192, 195
    A. radiobacter 193, 195
Alkaline phosphatase 82, 83, 94, 96, 110, 160,
    183, 205, 207
Aminoacid decarboxylase 69
Aminopeptidase 60
Amoxicillin with clavulanic acid 68
Ampicillin 68
Anaerobic cultivation 47–50, 62, 115, 131, 132,
    140, 146, 170, 204, 206, 210
Anaerobic growth 55, 62
Anatomy 2
Anesthesia 19, 20, 121, 179
Anesthetics 49
AniCard 204
Anthrax 141
Anti-RNA immunoassay 134
Antibiotic sensitivity 68
Antibiotic treatment 65, 66, 143, 145, 168
Antibiotic-associated colitis 143
Antibiotics 49
Antigen 75–77, 81, 83, 86, 87, 90, 91, 93–95, 97,
    98, 121, 122, 160, 198, 200, 211, 212,
    217, 219, 220, 228, 229
Antigen boiling 93
Antigen detection 75
Antigen-coated slides 82
Antisera 49, 81, 219
APAAP staining 83
API 20 ANAEROB 124
API 20 Strep 122–124, 131
API 20A 131, 139, 141, 147, 204
API 20E 142, 171, 195

245

API 20NE 178, 180, 185, 192, 195, 197, 198
API 50CHL 131, 142
API Campy 208, 210
API Coryne 131, 134, 138, 139
API Listeria 134
API NH 182, 183
API STAPH 114, 118
API system 69, 70
Arcanobacterium 129, 134, 136, 138
    A. haemolyticum 136, 138
Arthritis 120, 131, 137, 184, 224
ASM's *Manual of Clinical Microbiology* 60
Association 3
Association flora 130
ATCC 1470 modified leptospira medium 219
Atrophic rhinitis 177
Avidin 83

B

β-lymphocytes 131
Bacillus 129, 141–144, 228
    B. anthracis 141, 142, 144
    B. cereus 142–144
    B. circulans 144
    B. coagulans 144
    B. licheniformis 144
    B. macerans 144
    B. megaterium 144
    B. mycoides 143, 144
    B. piliformis 158
    B. polymyxa 144
    B. pumilus 144
    B. sphaericus 144
    B. subtilis 144
    B. thuringiensis 144
Bacillus spp. 61
BACTEC AFP 156
Bacterial interference 1
Bacteremia 5
Bacteroidaceae 59, 204
Bacteroides 48, 49, 204, 205
    B. gracilis 205
    B. ureolyticus group 205
    B. fragilis 205
Bacteriophages 118
Baird Parkers agar 117
Barbiturates 19
Basic characteristics (isolates) 55, 56
Basic tests 18, 55, 76, 128, 192
BB rats 224
Bergey's Manual 60, 131
Bile 69, 122

Biosafety levels 156, 176, 200, 209, 220, 239–244
Biotin 83, 110, 160, 212
Blocking 87, 94
Blocking buffer 95
Blood agar 43–45, 47, 48, 116, 120–122, 128, 130, 131, 133, 134, 137, 138, 145, 170, 177, 178, 180, 184, 195, 197, 206
Blood tellurite agar 137
Bordetella 192, 197
    B. avium 197
    B. bronchiseptica 4, 7, 23, 35, 43–45, 192, 193, 197
    B. parapertussis 197
    B. pertussis 197
Borrelia 215
Bound coagulase test 118
Bovine serum albumin 94
BPLS agar 43–45, 170, 171
Brain Heart Infusion medium (broth) 131
Bronchopneumonia 177, 197, 224
Brucella 197
Buffer tank blotting technique 98, 99
Burkholderia 59, 192, 195
    B. cepacia 193, 195

C

C57BL mice 154, 184, 209
Cecum 34, 35, 37, 43–45, 49, 51–53, 134, 138, 142, 158, 167, 171, 185, 195, 204, 210
CAMP test 69
Campylobacter 23, 35, 59, 206–209, 239, 242
    C. coli 56, 206, 207, 209
    C. jejuni 206, 207, 209
Campyslide 208
Capture technique 87
CAR bacillus 4, 7, 43, 76, 87, 211, 223
Carbohydrate assimilation 142
Carbohydrate fermentation 64, 67
Carbohydrate utilization 55, 64, 166
Carbon dioxide 15, 49, 121
Carbonate buffer 94, 95
Carcass 21
Cardiolipin test 217
Castration 4
Catalase test 55, 63, 114, 156
CCEY agar 145, 146
Cell cultures 159
Centers for Disease Control and Prevention 156, 176, 200, 209, 220, 239
Cephalosporins 68
Chemical carcinogenesis 167
Chemiluminescent detection technique 98

Chimp liver cell cultures 159
Chinese letters 137
Chloral hydrate 43–45
Chloramphenicol 68
Chocolate agar 43–45, 47, 48, 121, 128, 130, 131, 134, 140, 145, 146, 178, 180, 181, 184, 195, 197, 200, 206, 210
Cholesterol free media 223
Chronic respiratory disease 211
Chryseomonas 59, 192, 195
    C. luteola 193, 195
Citrate buffer 95
Citrate utilization 69
Citrobacter 4, 167, 169, 172, 173
    C. diversus 167, 173
    C. freundii 167, 172
    C. freundii type 4280 167
    C. freundii type ANL 167
    C. freundii type Ediger 167
    C. rodentium 4, 5, 7, 11, 23, 35, 43, 71, 167, 171, 172
Clindamycin 143, 181
Clostridium 129, 141, 143, 146–148, 228
    C. bifermentans 148
    C. botulinum 148
    C. butyricum 147, 148
    C. cadaveris 149
    C. chauvoei 148
    C. clostridioforme 145, 147, 148
    C. difficile 145–148
    C. histolyticum 148
    C. innocuum 149
    C. limosum 148
    C. novyi 148
    C. paraputrificum 149
    C. perfringens 145–148
    C. piliforme 4–7, 10, 11, 43–45, 56, 76, 77, 81, 82, 87, 102, 110, 141, 145, 153, 158, 159
    C. ramosum 149
    C. septicum 147, 148
    C. sordellii 148
    C. sphenoides 145, 149
    C. spiroforme 56, 145–147
    C. sporogenes 148
    C. subterminale 148
    C. tertium 149
    C. tetanii 149
    Toxins 145, 147
Clumping factor 118
Coagulase 69
Coagulase test 116, 118
Coagulum 20, 21
Coating 94
Coating buffer 93–95

Colibacillosis 167
Colistin 206
Columbia blood agar 121
Commensal 3
Commercial test kits 68
Complement fixation test 181, 229
Confidence limit 8, 10
Conjunctiva 23, 27, 180, 218, 224, 227
Conjunctivitis 120, 137, 177, 179, 192, 224
Contamination 2, 110
Control group 1
Control sera 95
Copathogen 3
Coprococcus 124
Coronavirus 145
Corynebacterium 129, 134, 135, 154
    C. bovis 136–138
    C. equi 139
    C. haemolyticum 138
    C. hoffmanii 137
    C. kutscheri 5, 7, 23, 35, 56, 135, 137, 138, 169
    C. minutissimum 136–138
    C. pseudodiphtericum 136, 137
    C. pyogenes 140
    C. renale 135, 137
    C. urealyticum 136–138
    C. xerosis 135, 138
Coryneform bacteria 134, 135
Cow milk 94
Cowan and Steel's Bacteriology 60, 69
Cristispira 215
Cryptosporidia 145
Cultivation 16
Cut-off value 68, 98
Cystitis 167
Cytochrome oxidase test 55, 64, 65, 181

D

Dark-field microscopy 216–220, 225
Day-to-day variation 97
Defined flora 204
Dehydration 145, 167
Deinococcaceae 114
Deinococcus 114
Deoxyribonuclease activity 69
Depression 120, 132
Dermal necrosis 224
Dermatitis 116, 137, 140
Desulfotomaculum 141
Detergents 49, 94
Determinant 3
Diarrhea 140, 145, 146, 167–169

Dichotomic identification 11
Diffusion inhibition assay 65, 67, 68, 228
Dilution buffer 94, 95
Dioxigenin 110
Disc methods 65
Diseased animals 11, 23, 35, 118, 138, 140, 146,
        156, 178, 184, 194
Disinfection 21
DNA 101, 105–107, 109, 110, 176, 228
Dormicum® 18, 19
Dot blotting 98, 105, 178, 228
Double strips 87
Dulbecco's medium 212
Dyspnoea 120

E

Eagle's minimum essential medium 212
Edema 141, 216
Edwardsiella 169, 173
        E. tarda 173
Egg Yolk Agar 147
Eimeria 145
ELISA reader 93
Embryonated eggs 159, 212
Encephalitis 133
Endocarditis 131
Endogenous peroxidase 83
Enhanced binding 87
Enriched media 118
Enrofloxacin 68
Enteritis 5, 145, 158, 167
Enterobacter 168, 172, 173, 175
        E. aerogenes 168, 172
        E. agglomerans 168, 172, 173, 175
        E. cloacae 5, 168, 169, 172
        E. gergoviae 168, 172, 173
        E. sakazakii 168, 172
Enterobacteriaceae 47, 58, 166, 169, 172, 173,
        175
Enterococcus 67, 115, 121, 123
        E. faecalis 122, 123
        E. faecium 122, 123
Enteropathies 167
Enterotube II 171
Environment 2, 120
Enzyme-antienzyme technique 83
        staining 212
Enzyme-conjugated antibodies 82, 93, 96, 98
Enzyme-linked immunosorbent assay 43–45,
        83, 86, 89, 91, 95, 98, 134, 138, 147,
        160, 178, 181, 182, 184, 195, 200, 211,
        212, 217, 229
Enzyme labeled probes 105

Erosions 216
Erysipeloid 131
Erysipelothrix 57, 129–131, 138
        E. rhusiopathiae 130–132
        E. tonsillarum 131
Erythrocyte adsorption 228
Erythromycin 68
Escherichia 166, 167
        E. coli 4, 35, 145, 167, 169, 173
        toxins 167
Ethanol 20–22, 62, 66, 146
Ethidium bromide 109, 110, 176
        staining 110
Eubacterium 129, 134, 139
        E. lentum 140
        E. limosum 140
Euthanasia 15
Experimental immunosuppression 4

F

F. tularensis 200
False negatives 8, 9
False positives 8, 82
Feces 23, 137, 146, 156, 176, 208
Federation of European Laboratory Animal
        Science Associations 6, 120, 171,
        132
Feeding 167
Fentanyl-fluanisone 18, 19
Fetal calf serum 94
Fever 166, 182, 218, 220
Flavobacterium 59, 198, 199, 211
        F. breve 198, 199
        F. indologenes 198, 199
        F. meningosepticum 198, 199
        F. odoratum 198, 199
Fluorescein 194
Fluorescein-conjugated, species-specific anti-
        immunoglobulins 77, 98
Francisella 59, 198, 200
        F. tularensis 200, 239, 242, 244
Fried-egg-type colonies 184, 225
Fucidine 68
Fusobacterium 204
        F. mortiferum 205
        F. necrophorum 204, 205
        F. necrophorum subsp. *funduliforme* 206
        F. necrophorum subsp. *necrophorum* 206
        F. nucleatum 205
        F. varium 205

G

Gas chromatography 147
Gas production 64
Gelatine liquefaction 69
Gemella 57, 115, 118, 123
    G. haemolysans 124
    G. morbillorum 124
Gene sequences 101, 102, 211, 228
Genetics 2, 4, 137, 158, 167, 184, 224
Genitals 35, 38, 39, 42, 43–45, 49, 53, 118, 134,
    138, 167, 171, 177, 178, 184, 185, 195,
    198, 216, 224, 227
Gentamicin 68, 143, 220
Gerbils 19, 44, 94, 116, 158, 179, 195, 197
Germ-free animals 95
Giemsa staining 160, 216–218, 225
Glucose fermentation 55
Gram staining 60, 62, 160, 225
Gram-stainability 55, 60
Granulomas 154, 155
Guinea pigs 4, 15, 18, 19, 34, 45, 120, 132, 137,
    143, 145, 155, 156, 167–169, 182, 184,
    192, 195, 197, 198, 223, 224, 227, 229
Gyrotory water bath shaker 210

H

H_2S production 130, 138
Haemophilus 43, 47, 64, 177, 181–183
    H. influenza murium 181, 182
Hafnia 169, 172
    H. alvei 172
Hamsters 19, 44, 137, 140, 143, 155, 158, 167,
    197, 200, 206, 209, 227, 229
Health status 3
Healthy animals 5, 23, 35, 118, 131, 134, 138,
    140, 146, 156, 158, 170, 178, 184, 206,
    217
Heart puncture 20
Helicobacter 7, 35, 59, 206–211
    H. bilis 207, 209–211
    H. cholecystus 209
    H. cinaedi 207, 209
    H. hepaticus 43, 56, 87, 107, 207, 209–211
    H. muridarum 207, 209–211
    H. rappini 207, 209–211
    H. rodentium 207, 209
    H. trogontum 207, 209, 210
hemolysins
    α 116
    β 116
    γ 116
    δ 116

Hemolysis 19, 47, 116, 121, 145, 200
Hepatitis 137, 209, 211
High binding plates 87
High agar 62, 65
High-performance liquid chromatography
    156
Horse serum 225
Hugh and Leifson's medium 65
Humane immunedeficiency virus 158
Humid chamber 80
Hydrogen sulphide production 69
Hypnorm® 18, 19

I

Identification 55
IgG 76, 229
IgM 229
Immune-competent animals 116, 137, 192
Immune-deficient animals 4, 116, 123, 192,
    194, 224
Immunoblotting 98
Immunocard® 147
Immunodiffusion 160
Immunoenzymatic staining 82, 83
Immunofluorescence techniques 77, 81, 82
    immunofluorescence assay 45, 81, 160,
    182, 212, 217, 229
    indirect technique 81
    staining 77, 79, 160, 212, 216–218, 225, 228
Immunological methods 75
Immunology 1, 2
Immunosuppressed animals 5, 10, 116, 192
Immunosuppression 4, 5, 10, 116, 137, 179
In situ hybridization 105
Inapparent carrier 3
Inbred strains 4, 158
Incidence 3
Incubation 44, 45, 49, 51, 53, 59, 64, 67, 95, 114,
    116, 121, 123, 131, 137–140, 142,
    145–147, 166, 169, 170, 180, 181, 195,
    208, 210, 226
Incubation temperature 95
India ink staining 216
Indicative media 49, 53
Individually ventilated cages 10
Indole reaction 69
Inducer 3
Infection 2, 3
    dormant subclinical infection 3
    latent subclinical infection 3
    subclinical infection 3
Infective dose 4
Infectivity 3

Instruments 20, 22
Intestinal disease 116
Isolation 16, 51, 53
    grouped animal/grouped organ 53
    individual animal/individual organ 51
    grouped animal/individual organ 52
    individual animal/grouped organ 52
Isolators 10

J

Jaundice 218

K

Kanamycin 206
KCN resistance 171–173, 175
Ketalar® 19
Ketamine 19
Ketaminol® 19
Ketaset® 19
Kidneys 36, 192, 218
King's agars 194
Klebsiella 4, 168, 175
    K. orthinolytica 168, 175
    K. oxytoca 168, 175
    K. planticola 168, 175
    K. pneumoniae 5, 35, 43–45, 168, 169, 175
Klyuvera 169, 172
Kovács' reagent 65
Kurthia 57, 128, 129
    K. gibsonii 128
    K. zopfii 128

L

L-929 cell line 159
Lactobacillus 4, 129, 130, 138, 228
    L. lactis lactis 131
Lactose-sucrose agar 197
Lancefield's groups 76, 118, 121–123
Latex particles 76
Lecithinase 69, 147
Lepra cells 154
Leptospira 7, 43, 76, 215, 218, 239, 242
    L. interrogans 218–220
    L. interrogans serovar *icterohemorrhagiae* 218, 220
    L. interrogans serovar *ballum* 218, 220
    L. interrogans serovar *canicola* 220
Leptospiraceae 215
Leptospirosis 218
Lethal dose 4

Leuconostoc 118
Lincomycin 68, 143, 181
Lipase 69, 147
Liquefactive necrosis 140
Listeria 32, 87, 105, 129, 130, 132, 133, 135
    L. grayi 135
    L. innocua 135
    L. ivanovii 135
    L. monocytogenes 132–135
    L. murrayi 135
    L. seeligeri 135
    L. welshimeri 135
Listeria agar 133
Listeria Enrichment Broth 133
Liver 5, 35, 36, 40, 41, 48, 120, 132, 137, 158–160, 192, 206, 209–211
    necrosis 5, 158
    tumors 209
Lungs 35, 37, 154, 155, 178, 180–182, 197, 212
Lymph node swelling 182
Lymphadenitis 120
Lysis buffer 109
Lysostaphin 114, 115
Löwenstein-Jensen slants 156, 157

M

MacConkey agar 138, 156, 170, 177, 184, 195, 197, 198
Maldigestion 145
Malonate utilization 69
Mannitol salt agar 117
Mastitis 120, 167, 179, 180
Maxisorp 87
Media 47
Medium binding plates 87
Megaloileitis 158
Meningitis 133
Meritec Campy Jcl 209
Metritis 167
Mice 4, 19, 43, 116, 120, 132, 137, 141, 145, 154–156, 158, 159, 167, 168, 179, 181, 182, 192, 197, 206, 209, 211, 224, 227
Micro-ID 171
Microaerophilic cultivation 43–45, 49, 50, 121, 130, 131, 133, 170, 173, 181, 184, 208, 210, 226
Microbiological entity 10
Microbiological status 6
Microbiologically defined animals 5
Micrococcaceae 47, 113, 114
Micrococcus 57, 114, 115
    M. kristinae 115
    M. luteus 115

M. roseus 115
M. sedentarius 115
M. varians 115
Microplate washer 92
Microscopy 56, 60, 82, 130, 138, 216, 218
Microtiter plates 87, 97
    polystyrene 87
    PVC 87
Midazolam 18, 19
Middle ear 23, 27, 28, 192
Minitek 204
Mismatching 108, 109
Molecular biological methods 101, 160, 181,
    211, 217
Molecular probes 105, 107, 110, 118, 122, 134,
    156, 220
Mollicutes 223
Monofactorial disease 3
Morganella 166, 168, 172
    M. morganii 4, 168, 169, 172
Motility 55, 57, 58, 62–64, 132, 144, 166,
    171–173, 175, 192, 216, 217, 225
Mucopurulent discharge 120
Mueller Hinton agar 67, 68, 181
Multifactorial disease 3
Multiple pyogranumalotous inflammation
    140
Murine leprosy 154
Murine leukemia cell lines 224
Mycobacterium 23, 35, 105, 153, 154, 156, 157,
    239, 242
    M. avium 155, 156
    M. avium intracellulare 154, 155
    M. bovis 155, 156, 242, 244
    M. lepraemurium 154, 155
    M. microti 154, 155
    M. tuberculosis 155, 156, 242
Mycoplasma 7, 23, 35, 76, 81, 87, 110, 212, 223,
    226–228
    M. arthritidis 224, 225, 227–229
    M. caviae 224, 227
    M. cavipharyngis 224, 227, 228
    M. collis 224, 227, 228
    M. cricetuli 224, 227, 228
    M. muris 224, 225, 227, 228
    M. neurolyticum 224, 225, 227, 228
    M. neurolyticum toxin 224
    M. oxoniensis 224, 227, 228
    M. pulmonis 2, 4, 43, 179, 223, 225, 227, 228
Mycoplasma agar 226
Mycoplasma broth 226
Mycoplasmosis 4
Myocardial necrosis 158

N

Narcoxyl® 19
National Institutes of Health 6
NCTC 1469 cell line 159
Necrosis 5, 140, 158, 204, 224
Neomycin 68, 219
Neonatal diarrhoea 5
Nephritis 137, 167, 224
Neufeld reaction 121
NH_3 2
Nicotine amide dinucleotide 181
Nitrate reduction 69
Nitrocellulose membranes 98, 110
Nocardia 129, 140, 154
    N. asteroides 140
Non-immunoglobulin binding protein 94
Non-persisting infections 107
Non-selective media 16, 47, 49, 55
Norleucine-tyrosine-broth-culture 147
Nose 23, 26, 34, 43–45, 51–53, 121, 131, 132,
    140, 178, 184, 197
Nucleases 105, 108
Nude mice 4, 116, 137
Nutritional deficiencies 120
Nylon membranes 98, 105, 110

O

O-nitro-phenylgalactoside 96
O-phenylenediamine 96
O/129 185
OD-value 86, 97, 98
Oerskovia 129, 134, 136, 139
    O. turbata 136, 139
    O. xanthineolytica 136, 139
Oncology 1
ONPG test 69
Opportunist 3
Opthalmia 120
Opthalmitis 179
Optochin sensitivity 122
Otitis interna 120
Otitis media 120, 177, 179
Oxalic acid method 156, 157
Oxidase test 192
Oxygen 15

P

P-nitrophenylphosphate 96
PALCAM agar 133
Pancreatitis 209

PAP staining 83, 85
Parainfluenza virus type III 197
Pasteurella 7, 43–45, 58, 177, 179
    P. dagmatis 180
    P. haemolytica 179, 180
    P. multocida 177, 178
    P. multocida subsp. *gallicida* 178, 179
    P. multocida subsp. *multocida* 178, 179
    P. multocida subsp. *septica* 178, 179
    P. pneumotropica 4, 47, 61, 69, 87, 103, 109,
        177–182, 223
    P. pneumotropica type Henriksen 180
    P. pneumotropica type Heyl 178, 180
    P. pneumotropica type Jawetz 178, 180
    P. ureae 180
Pasteurellaceae 23, 35, 43, 47, 109, 166, 177,
    178, 181, 182
Pasteurellosis 179
Pathogen 3
Pathogenicity 1, 3
Pathogenic 3
Pediococcus 57, 118, 123, 124
Penicillin 68, 220, 225
Pentobarbitone 15, 19
Peptococcaceae 114, 124
Peptococcus 124
Peptostreptococcus 124
Peribronchial lymphoid cuffing 211
Pericarditis 120, 197
Periorbital puncture 20
Peritonitis 120
Peroxidase 82, 83, 94, 96, 110, 176
Peroxidase-antiperoxidase staining 77, 160
Persisting infections 2, 107
Petechial exanthema 182
Phage typing 118
Phase-contrast microscopy 216, 218, 225
Phenylalanine deaminase 69
Phosphatase 160, 176
Physiology 1
Pin-point necroses 132
Planning 15, 16
Planococcus 114
Plasmides 110
Platin needle 20
Pleuritis 120, 197
Pneumonia 116, 120, 140, 167, 168, 177, 181,
    185, 192, 197, 224
PNPX test 69
Polyarthritis 182
Polymerase chain reaction 105, 107, 131, 147,
    156, 176, 178, 181, 182, 200, 211, 212,
    217, 218, 228
    detection 109
Polymerase inhibitors 109

Polymerases 105, 108
    AmpliTaq 108
    Pfu 108
    rTth 108
    Taq 108
    Vent 108
Polysorp 87
Polystyrene tray 22
Polyvinylidene chloride membranes 98
Porphyrin formation 69
Porphyromonas
    P. asaccharolytica 205
    P. gingivalis 205
Post-surgical infections 4
Potassium cyanide resistance 69
Potassium hydroxide assay 60, 63
Prednisolone 159
Presumpto plate system 147
Prevalence 7, 9
Prevotella 205
    P. intermedia 205
Primary cultures 16, 18
Primers 105, 109, 228
Probiotics 4, 123, 130
Profiling 11, 47, 56
Protector 3
Protein analysis 160
Proteinases 109
Proteus 49, 166, 168, 170–172
    P. mirabilis 168, 169, 172
Protohemin 181
Protoporphyrin 181
Providencia 169, 175
    P. rettgeri 175
Pseudomonas 23, 35, 59, 171, 192–194, 196
    P. aeruginosa 2, 4, 5, 192, 194, 196
    P. alcaligenes 196
    P. cepacia 195
    P. diminuta 192, 196
    P. fluorescens 195, 196
    P. mendocina 196
    P. paucimobilis 195
    P. pseudoalcaligenes 196
    P. putida 194, 196
    P. stutzeri 196
    P. vesicularis 196
Pseudotuberculosis 5, 169
Pseudovertical contamination 179
Puff balls 184
Pulmonary edema 5
Pulmonary emboli 137
Pure cultures 11, 18, 51, 55, 76, 109
Pyelonephritis 167, 224
Pyocyanin 194
Pyrococcus furiosus 108

Q

Quick-staining kits 83

R

Rabbit syphilis 216
Rabbits 18, 19, 34, 45, 114, 115, 119, 120,
    122–124, 129, 132, 134, 135,
    138–140, 143, 145, 154, 155, 158, 168,
    169, 172, 175, 177, 178, 180, 182, 195,
    197, 198, 200, 204–206, 211, 216, 217,
    223, 224, 227, 229
Radionucleotide-labeled antibodies 98
Rales 211
Random sampling 6
RAPID ID 32A 204
Rat bite fever 166, 182, 220
Rats 4, 19, 43, 120, 124, 132, 137, 155, 158, 167,
    168, 179, 181, 182, 184, 192, 195, 197,
    198, 206, 209, 211, 212, 218, 224, 227
Reaction buffer 108
Rectal prolapses 167
Reproduction 1
Respiratory disease 2, 4, 5, 120, 177, 179, 211
Respiratory system 118, 120, 121, 123, 138,
    181, 195, 198, 227
Restriction analysis 110, 228
Restriction endonucleases 110
Restriction enzyme analysis 211
Reverse transcription polymerase chain
    reaction 108, 110, 160, 228
Rhinitis 177, 179, 192
Rhodococcus 58, 129, 134, 136, 139, 154
    R. equi 135, 136, 139
RNA 101, 107, 109
Rodents 15, 18, 19, 34, 114–116, 119, 122–124,
    129, 131, 134, 135, 137–140, 154,
    167–169, 172, 173, 175, 179, 182, 195,
    198, 200, 204–206, 210, 223, 228
Rolling disease 224
Rompun® 19
Rotavirus 145, 167
Rough hair coat 211
Ruffled fur 145, 224
Ruminococcus 124

S

Salmonella 5, 7, 11, 23, 35, 43, 45, 87, 166,
    168–171, 173, 176, 239, 242
    antigens 176

S. *enteritidis* 168
S. *typhimurium* 168
Salmonellosis 5, 168
Sample size 7, 9, 10
Sampling 6, 7, 18, 21, 23, 121
Sampling frequency 9
Sandwich technique 86, 89, 91
Saprophytic 3
Sarcina 124
Scandinavian Society for Laboratory Animal
    Science (Scand-LAS) 6
Screening 11
Selective media 16, 49
Selenite broth 43–45, 170, 176
Semi-solid media 62–64
Sendai virus 179
Sensitivity 8, 9, 82, 97, 105, 181, 195
Septicemia 120, 131, 132, 141, 168, 169, 177,
    178, 184, 192
Serology 18, 75, 76, 86, 87, 98, 156, 197, 217,
    228
Serratia 169
    S. *liquefaciens* 172
    S. *marquescens* 172
Serum agar 39, 40, 184
Serum broth 184
Sex 4
Sexual cycle 4
Shape 55
Short product 106
Silver stains 212
Single strips 87
Sinuitis 179
Skin 23, 25, 118
Skin lesions 131
Snuffles 177, 224
Sodiumdodecyle sulphate electrophoresis 98
Sodoku 220
Solid-phase hybridization 105
Solution-phase-hybridization 105
Sonication 93
Soy broth 121
SP4 agar 226
SP4 medium 225
Specificity 9, 82, 108, 110
Spectinomycin 68
Sphingomonas 59, 195
    S. *paucimobilis* 193, 195
Spiramycin 68
Spirillum minus 182, 216, 220
Spirochaeta 215
Spirochaetaceae 215
Spirochetes 215
Spiroplasma 228
Spleen 36, 120, 132, 192

Splenomegaly 5
Spore staining 64, 66
Spores 36, 55, 57, 64, 66, 132, 140, 142, 143,
        145, 147, 159, 212
Sporolactobacillus 58, 129, 141
Sporulation medium 142
Staphylococcal disease 116
Staphylococcus 5, 23, 57, 67, 114, 115
        S. aureus 4, 56, 76, 114, 116, 118, 119, 135
        S. cohnii 116, 119
        S. haemolyticus 116, 119
        S. hominis 119
        S. hyicus 118
        S. lugdunensis 119
        S. saphrophyticus 119
        S. sciuri 116, 119
        S. simulans 119
        S. xylosus 4, 116, 119
Stenotrophomonas
        S. maltophilia 195
Sterility 11, 15, 20, 21
Stomatococcus 114
Streaking 25
Streptavidin 83, 110, 212
Streptobacillus moniliformis 5, 7, 11, 23, 35, 43,
        44, 59, 166, 182, 184, 185, 220
Streptococcaceae 47, 113, 118, 122
Streptococcus 4, 57, 76, 105, 114, 115, 118, 228
        α-hemolytic 120
        β-hemolytic 120
        β-hemolytic streptococci 7, 23, 35, 43–45
        group A 105, 118, 120, 122
        group B 118, 120–122
        group C 4, 118, 120–122
        group F 121
        group G 118, 120–122
        non-hemolytic 120
        S. agalactiae 122
        S. dysgalactiae 122
        S. equisimilis 122
        S. morbillorum 124
        S. pneumoniae 4, 5, 7, 23, 35, 43–45, 56,
                120–122
        S. pyogenes 122
        S. zooepidemicus 56, 120, 122
Streptomyces 129, 140
Streptomycin 68
Stress 2, 4, 116, 120, 121, 179, 192
Stress testing 10
Subcultivation 16
Submandibular abscesses 120
Subserosal hemorrhage 218
Substrate 83, 94, 96
Sulphadoxine 159
Sulphonamide 68

Sulphuric acid method 156, 157
Swarming 49
Symbiont 3

T

Tartrate utilization 69
Teflon-coated slides 80, 81
Tellurite resistance 69
Tellurite-containing agars 123
Template DNA 109
Test group 1
Tetracyclines 68, 143
Tetramethyl-benzidine hydrochloride 96
Thermocycler 105
Thermus aquaticus 108
Thermus thermophilus 108
Thorax 33, 34
Tiamuline 68
Toxin neutralization assays 147
Trachea 15, 23, 29, 30, 34, 43–45, 51–53, 121,
                134, 142, 178, 180, 182, 184, 197, 212
Transplacental infection 137, 159
Treponema 215, 216
        T. cuniculi 41, 82, 216, 217
        T. pallidum 216, 217
Trimethoprim 68, 158, 159
Triple Sugar Iron Slant 130
True negatives 8
True positives 8
Tryptophanase 69
Tuberculosis 154
TVP agar 39, 208, 210
Tylosin 68
Tyzzer's disease 4, 5, 158

U

Ulcers 5, 200, 216
Unexpected deaths 120
Unspecific binding 94
Ureaplasma 228
Urease 69, 94
Urease test 206
Urinary calculus 137
Uveitis 224

V

V-factor 181, 182
Vancomycin 206
Vesicles 216
Vetalar® 19

Vibrio 185
Vibrionaceae 166, 185
Virulence 3
Vitamin deficiencies 2, 4
Voges-Proskauer test 69, 138

W

Warfarin 159
Warthin-Starry staining 212
Washing buffer 95
Wasting disease 169
Weeksella 198, 199
    W. virosa 198, 199
    W. zoohelcum 198, 199
Weight loss 132, 140, 211
Western immunoblotting 98, 160, 228

Wheezing 211
Working plan 16
Wound infection 182, 185, 198

X

X-factor 181, 182
Xanthomonas 59, 192, 195
    X. maltophilia 193, 195
Xylazine 19

Y

Yersinia 11, 43–45, 166, 169, 170, 175
    Y. pseudotuberculosis 7, 23, 35, 169, 175